The Art of Resisting Extreme Natural Forces

WIT*PRESS*

WIT Press publishes leading books in Science and Technology.
Visit our website for new and current list of titles.
www.witpress.com

WIT*eLibrary*

Home of the Transactions of the Wessex Institute.
Papers presented at ENGINEERING NATURE 2007 are archived in the WIT eLibrary in volume 58 of WIT Transactions on Engineering Sciences (ISSN: 1746-4471).
The WIT eLibrary provides the international scientific community with immediate and permanent access to individual papers presented at WIT conferences.
Visit the WIT eLibrary at www.witpress.com.

First International Conference on
the Art of Resisting Extreme Natural Forces

ENGINEERING NATURE 2007

Conference Chairmen

S. Hernández
University of La Coruña, Spain

C. A. Brebbia
Wessex Institute of Technology, UK

International Scientific Advisory Committee

Organised by
Wessex Institute of Technology, UK
University of La Coruña, Spain

Sponsored by
WIT Transactions on Engineering Sciences

WIT Transactions on Engineering Sciences

Transactions Editor

Editorial Board

The Art of Resisting Extreme Natural Forces

Editors

S. Hernández
University of La Coruña, Spain

C. A. Brebbia
Wessex Institute of Technology, UK

Editors:

S. Hernández
University of La Coruña, Spain

C. A. Brebbia
Wessex Institute of Technology, UK

Published by
WIT Press
Ashurst Lodge, Ashurst, Southampton, SO40 7AA, UK
Tel: 44 (0) 238 029 3223; Fax: 44 (0) 238 029 2853
E-Mail: witpress@witpress.com
http://www.witpress.com

For USA, Canada and Mexico

Computational Mechanics Inc
25 Bridge Street, Billerica, MA 01821, USA
Tel: 978 667 5841; Fax: 978 667 7582
E-Mail: infousa@witpress.com
http://www.witpress.com

British Library Cataloguing-in-Publication Data

A Catalogue record for this book is available
from the British Library

ISBN: 978-1-84564-086-6
ISSN: 1746-4471 (print)
ISSN: 1743-3533 (on-line)

The texts of the papers in this volume were set individually by the authors or under their supervision. Only minor corrections to the text may have been carried out by the publisher.

No responsibility is assumed by the Publisher, the Editors and Authors for any injury and/or damage to persons or property as a matter of products liability, negligence or otherwise, or from any use or operation of any methods, products, instructions or ideas contained in the material herein. The Publisher does not necessarily endorse the ideas held, or views expressed by the Editors or Authors of the material contained in its publications.

Printed in Great Britain by Cambridge Printing

Preface

This book contains papers presented at the first International Seminar on the Art of Resisting Extreme Natural Forces, held at the Wessex Institute of Technology in 2007.

According to the ancient Greeks, nature was composed of four elements: air, fire, water and earth. Engineers are continuously faced with the challenge imposed by those elements, in activities such as the design of bridges and high buildings to withstand high winds; the construction of fire resistant structures, the control of floods and wave forces; the minimisation of earthquake damage; prevention and control of landslides and a whole range of natural forces.

Advances in engineering design and the quest for optimal and more efficient structures have highlighted the importance of those natural forces with regard to the safety and durability of all forms of construction.

Natural disasters occurring in the last few years have stressed the need to achieve more accurate and safer design against extreme natural forces. At the same time, structural projects have become more challenging. Because of this, it is important to provide information on the current capabilities and problems in this field and the Editors hope that this book will help in this regard.

The Editors are grateful to all authors for their contribution as well as to the colleagues who helped to review them.

The Editors
The New Forest, UK
2007

Contents

Designing challenging bridges in northwest Spain

F. Nieto, S. Hernández, A. Baldomir & J. Á. Jurado
School of Civil Engineering, University of La Corunna, Spain

Abstract

This work presents two proposals for spanning the Galician rías. The rías are the equivalent to the Scottish firths or the Scandinavian fiords, thus long-span bridges are required to communicate both sides of that physical barrier. Additionally, the multidisciplinary approach applied in the design of these proposals that have been developed in the last decade by the Structural Mechanics Research Group is explained.
Keywords: suspension bridges, sensitivity analysis, optimum design, advanced visualization, aeroelasticity.

1 Introduction

Galicia is the name of the Spanish region located in the northwestern corner of the Iberian Peninsula (see figure 1). This region does not have high mountains nor large rivers, the main characteristic of the Galician physical geography is the existence of "rias". The rias are sea waters that have advanced deep into the land, in the mouths of some small rivers (see figure 2). Thus, there are two different ways to understand the rias as a frame for the territory: on one hand as sea masses that have penetrated into the land and on the other hand they are also peninsulas that have advanced into the Atlantic ocean. This fact has given birth to a very special and delicate environmental balance that hosts important industries related with the seafood. At the same time, due to their characteristics as natural harbours, they also host crucial industrial and commercial activities and, therefore, the two main metropolitan areas around the cities of La Corunna in the north and Vigo in the south (marked with yellow frames in figure 2) are located around the "Rias Altas" and the "Ría of Vigo", respectively. The existence of the rias has constrained both the road and railway networks in

www.witpress.com, ISSN 1743-3533 (on-line)
doi:10.2495/EN070011

Galicia throughout history. To date, the main communication route has followed the north-south orientation along the west coast, but always avoiding spanning the rias. The consequence has been the exclusion of the land peninsulas between the rias from the main communication networks which has been an important drawback in the social and economical development of the Galician west coast.

Figure 1: Map of Spain and Galicia location.

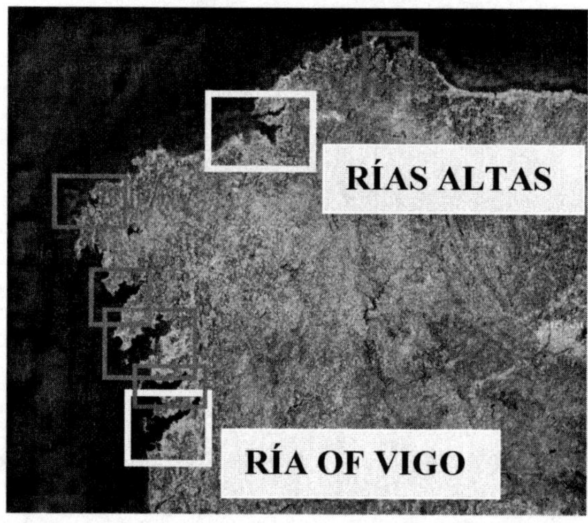

Figure 2: Map of Galicia and location of the main "rías".

However this geographic scenario is not something exclusive of this region. Other very similar landscapes are the Scottish "firths" and the Scandinavian "fiords". It is a wise exercise to analyze how different countries face their geographic constraints and how the provided solutions rely not only on the national economical power, but also on the envision of their engineers and both political and social leaders. Valuable lessons can be learned from the experiences of the United Kingdom and Norway.

Actually, in Scotland three bridges span the natural barrier of the Firth of Forth (figures 3 and 4): the Forth railway Bridge built in 1890, the Kincardine Bridge opened in 1936 and the Forth road bridge inaugurated in 1964. The

existence of these three bridges built along 75 years is a clue of the importance that some societies give to communication networks. In fact, currently a new road bridge is being constructed parallel to the existing Kincardine Bridge in order to cope with the existing traffic volume [1].

Figure 3: The Kincardine Bridge spanning the Firth of Forth.

Figure 4: The road and railway bridges over the Firth of Forth.

Another good example of how British people have tackled with transport difficulties is the construction of the Humber Bridge. "For a long time the Humber Estuary was a barrier to trade and development between the two banks and local interests campaigned for over 100 years for the construction of a bridge or tunnel across the estuary" [2]. Finally, the actual bridge was inaugurated in 1981. The opening of the bridge has improved communication enabling the area to realise its potential in commercial, industrial and tourist development. Additionally, it has saved millions of vehicle kilometres as well as drivers and passengers time.

Another good modern example of the aim for overcoming geographic barriers is the Triangle Link [3]. The target of this large and complex project has been to connect two large and several small islands to the main land of Western Norway by means of a 7,9 km tunnel and two suspension bridges: the Storda Bridge with a main span of 677 m and the Bømla Bridge with a main span of 577 m.

Figure 5: Triangle link location and its two suspension bridges.

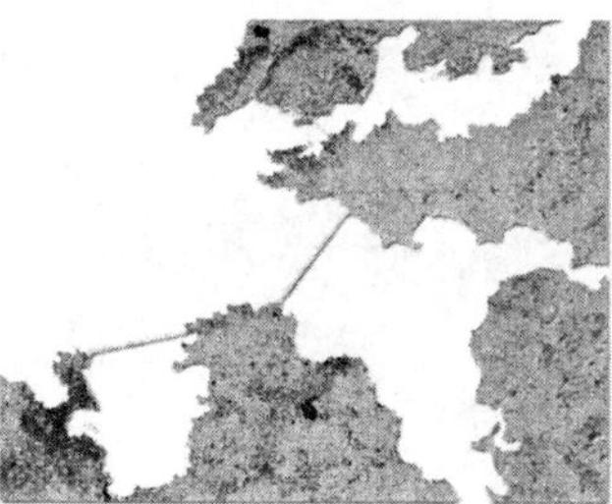

Figure 6: Location of the Rías Altas link.

2 Spanning the Galician Rías

On the wake of the existing successful routes all over the world, two challenging links have been proposed by the Structural Mechanics Research Group of the University of La Corunna for the Galician rías: the Rías Altas link in the North and the Balcony Bridge over the Ría of Vigo in the South.

2.1 The Rías Altas link

The Rías Altas link [4,5] is a project of a new infrastructure spanning three of the rías located in the North of Galicia. The link will connect the cities of La Corunna and Ferrol which are nowadays joined by a road and a toll highway along distances around 65 Km. The Rías Altas link would decrease that distance to 13.9 Km. Figure 6 shows a detailed map of the area where the Rias Altas link crossing is projected. Furthermore, in figure 7 the layout of the proposal is presented. It can be seen that two typologies of bridges have been proposed: suspension and arch bridges.

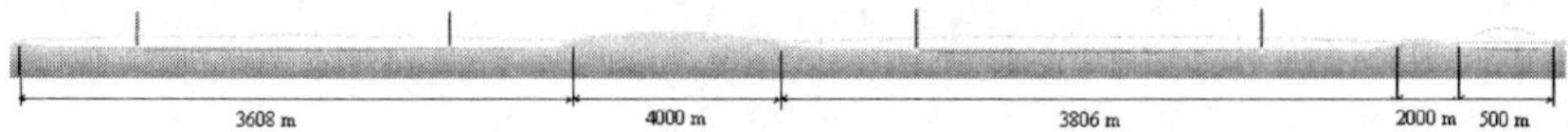

Figure 7: Layout of the Rías Altas link.

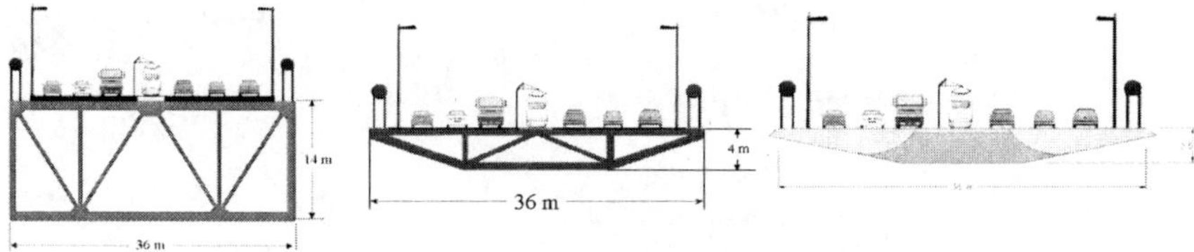

Figure 8: Deck cross-section conceptual design.

Figure 9: Digital picture of the Ares Bride.

Two suspension bridges have been included in this proposal. They are La Corunna and Ares Bridges. Both have very long main spans: 2016 m for La Corunna Bridge and 2198 m for the Ares Bridge. The towers highs are 260 m for La Corunna Bridge and 296 m for the Are Bridge.

Regarding the deck cross-section, three different concepts have been considered: a truss girder, a single aerodynamic box and two-box aerodynamic deck with an open grid between the two boxes. The conceptual design of all of them is presented in figure 8. Furthermore, a digital picture of one of the suspension bridges incorporating the existing landscape is shown in figure 9.

The third bridge included in the project is the Ferrol Bridge, corresponding to the northernmost position of the Rías Altas link. This is an arch bridge with a span length of 500 m. The deck is at an intermediate position with respect to the arch and it has an aerodynamic cross-section aiming to favour an efficient behaviour under wind-induced loads. Stiffening elements joining the twin arches of the bridge have the same pattern as the towers of the suspension bridges.

2.2 The Balcony Bridge over the Ría of Vigo

This bridge has been projected as a solution for the traffic congestion problem generated as the existing cable-stayed bridge over the Ría of Vigo is not able to cope with the current number of vehicles. The proposed alternative is a new suspension bridge downstream of the existing cable-stayed bridge.

The projected bridge has a main span of 1800 m and two secondary spans of 650 m each. The chosen deck cross-section is a 21.5 m wide and 4 m depth symmetric box. Special care has been taken in the aesthetics of the bridge.

One of the key issues of this proposal is the envision of the bridge receiving pedestrians as it will communicate two populated urban areas and the bridge surroundings will attract a number of visitors due to the wonderful existing landscape. In fact, balcony zones have been dedicated around the towers at deck level to allow the recreational use of the structure. In figures 10 and 11 digital visualizations of the Balcony Bridge over the Ría of Vigo are presented.

Figure 10: Balcony Bridge visualization.

Figure 11: Balcony area around the towers.

3 Analysis of suspension bridges: a multidisciplinary approach

The analysis of challenging structures as the ones presented in the former point involves a careful study in order to decide the acting loads and their distribution. In fact, many of the actions to be considered are natural loads in the sense that they are caused by environmental elements. Some of them are listed ahead:

- Wind-induced loads: they are responsible for phenomena such as: static loads acting on the bridge, flutter instability, buffeting, vortex shedding, cable vibrations or pedestrian discomfort amongst others.

- Earthquake solicitations: they depend on the existing earthquake risk in the bridge location and their expected intensity.
- Sea loads acting on the towers such as waves or the ones cause by tides.
- Thermal loads caused by the environment.
- Snow and ice on the bridge superstructure.

From the aforementioned loads, the ones induced by the wind are the most important in the design of suspension bridges. Moreover, prevention against the destructive aeroelastic phenomenon of flutter must be surveyed in depth as it represents one of the most relevant constraints for the whole design of the bridge.

The Structural Mechanics Research Group has worked extensively for years on defining feasible design strategies of long-span bridges. As a result, a multidisciplinary approach has been proposed and applied to a number of bridges. That methodology comprises aeroelastic studies, sensitivity analyses and optimum design techniques as well as advanced visualization.

3.1 Aeroelastic studies

Suspension bridges are wind prone structures. Wind-induced instabilities as flutter can be responsible for the bridge destruction. Thus, special care must be taken to achieve a feasible and safe aerodynamic design. To date, two reliable methods exist in order to analyze the bridge behaviour under wind loads. On one hand, full model testing in boundary layer wind tunnels allows the measurement of model structural responses and the flow speed leading to instabilities such as flutter of buffeting can be observed and evaluated. On the other hand, section model testing of bridge decks in wind tunnels under different flow speeds with the aim of identifying the aerodynamic coefficients and flutter derivatives [6] which are parameters employed in the computer evaluation of the bridge response under wind loads.

3.2 Sensitivity analysis and optimum design techniques

The design process of large and complex structures like the proposed suspension bridges for spanning the Galician rías requires a great number of intermediate candidate designs that are evaluated both experimentally and computationally. To date, each design modification is based upon the experience and expertise of the project team, but this circumstance does not guarantee that a modified design is going to perform better that the previous one, which can lead to an inefficient design process.

Alternative approaches are the ones based upon sensitivity analysis and optimum design, that have been extensively used in car or aircraft industries with very good results. Sensitivity analysis can be defined as a response derivative with regards to a design variable, that is, a structural property with a potential for change. This derivative can be understood as the expected change in the studied response when the considered design variable is perturbed. Therefore, the structural behaviour due to a change in a design variable can be anticipated. This fact allows the designer to follow a guided design process avoiding unfruitful design modifications. Practical applications of sensitivity analysis in the bridge engineering realm are limited [7–10].

A classical definition for optimum design is the one by Wilde [11]: the best feasible design according to a preselected quantitative measure of effectiveness. In optimal structural design a certain objective function, for instance the structure weight, must be minimized or maximized by modifying the design variables while satisfying a set of behaviour and design constraints. The final optimum design can be accomplished by mathematical methods thanks to the use of computers. Furthermore, when a gradient based method is implemented, evaluation of the sensitivities of the objective function and the constraints is entailed. Several examples of structural optimization can be found in the literature [12–14]. The authors have carried out the optimum design of the record-breaking Messina Strait suspension bridge [15,16].

3.3 Advanced visualization

A suspension bridge is a landmark, thus its design involves multiple considerations regarding aesthetics, serviceability, or aeroelastic performance, amongst others. Currently, both a visualization model which represents the appearance of the bridge and a finite element model created to perform the required structural analyses can be worked out. In civil engineering the use of digital models has been mainly devoted to anticipate the visual appearance of the projected structure, but they also are very appropriate for simulate dynamic responses as vibration modes or structural deflections of bridges under wind-induced instabilities as flutter. In that respect, many of the experiments carried out in boundary layer wind tunnels can be reproduced by means of computer animations as if it was a virtual wind tunnel. This methodology was first applied to a model of the Tacoma Narrows Bridge which collapsed in 1940 with the aim of simulating its behaviour until reaching the flutter critical wind speed [17].

The methodology carried out to visualize aeroelastic deformation of suspension bridges involves much kind of calculations and a quite large set of different softwares has been used. CIVISM, FLAS and ADISNOL codes have been developed in the University of La Coruña and MAYA [18] is a commercial code. A list of the required task with indication of the associated codes is as follows:

1) A set of data is introduced in the computer and the CIVISM code produces two types of files. One of them will be used as input for the structural model and the second one for the visualization model.

2) A file produced by CIVISM is read by the ADISNOL code and a structural model is created.

3) Calculation of eigenfrequencies and eigenvectors using second order theory is carried out using ADISNOL.

4) With data provided by ADISNOL an aeroelastic analysis of the suspension bridge using modal expansion of the displacements is worked out using the FLAS code. FLAS provides the deformed geometry of bridge for any specified wind speed.

5) A file produced by CIVISM is read by the visualization software MAYA to produce the visualization model.
6) Using in-house routines and the programming capabilities of MAYA the deformed geometry of the structural model obtained in step 4) is transferred to MAYA and the deformed geometry produced for a wind speed U at any specified time can be visualized.

Producing a large set of simple images, at a ratio of 25 frames per second, a computer animation can be recorded.

In figure 12 a diagram flow explaining the methodology is enclosed.

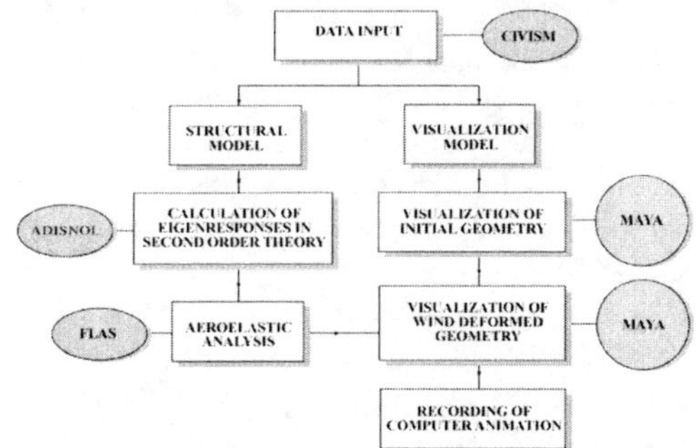

Figure 12: Flow char of the multidisciplinary design approach.

4 Conclusions

The Rías Altas link and the Balcony Bridge over the Ría of Vigo can be milestones in the social and economical development of the Spanish Northwestern region.

Multidisciplinary techniques involving advanced visualization, and both sensitivity analysis or optimum design methods can be of great significance in the feasible design of challenging structures in the civil engineering realm.

References

[1] Upper Forth Crossing at Kincardine http://www.upperforthcrossing.com/upper_forth/UF_FolderHomePage.jsp?pContentID=25&p_applic=CCC&p_service=Content.show&

[2] Humber Bridge web site http://www.humberbridge.co.uk/explore.php

[3] Schaathun, H. (2001) *The triangle link projects, design and construction.* Fourth Symposium on Strait Crossings 2001. (Jon Krokeborg Ed.). A.A. Balkena Publishers. Bergen, Norway, 2-5 September.

[4] Hernández S. & Sánchez A. (2000) *Era 2000. Proyecto de enlace en las Rías Altas.* University of La Corunna. (In Spanish).

[5] Hernández S. (2001) *The Rías Altas link. A challenging crossing.* Fourth Symposium on Strait Crossings 2001. (Jon Krokeborg Ed.). A.A. Balkena Publishers. Bergen, Norway, 2-5 September.

[6] Simiu, E. & Scanlan, R.H. (1996) *Wind Effects on Structures: Fundamentals and Applications to Design.* John Wiley & Sons.

[7] Wang, J., Ko, J. & Ni, Y. (2000) *Modal Sensitivity Analysis of Tsing Ma Bridge for Structural Damage Detection.* Nondestructive Evaluation of Highways, Utilities and Pipelines IV.

[8] Jurado J.A. & Hernández S. (2004) *Sensitivity Analysis of Bridge Flutter with respect to Mechanical Parameters of the Deck.* Structural and Multidisciplinary Optimization vol. 27(4), pp. 272-283.

[9] Nieto F., Hernández S. & Jurado J.A. (2005) *Distributed Computing for the Evaluation of the Aeroelastic Response and Sensitivity Analysis of Flutter Speed of the Messina Bridge.* Fluid Structure Interaction and Moving Boundary Problems, La Corunna, Spain, WIT Press.

[10] Nieto, F. & Hernández, S. (2006) *An Augmented Set of Design Variables for Sensitivity Analysis of Flutter Speed of Cable Supported Bridges.* 9° Convegno Nazionale di Ingegneria del Vento IN-VENTO 2006. Italy.

[11] Wilde, D.J. (1978) *Globally Optimal Design.* John Wiley & Sons, NY.

[12] Hernández, S. (1989) *Métodos de Diseño Óptimo de Estructuras.* Colegio de Ingenieros de Caminos, Canales y Puertos, Madrid. (In Spanish).

[13] Arora, J. (Ed.) (1997) *Guide to Structural Optimization.* Technical Committee on Optimal Structural Design, ASCE, New York.

[14] Burns, S.A. (Ed.) (2002) *Recent Advances in Optimal Structural Design.* Technical Committee on Optimal Structural Design, ASCE, Reston.

[15] Nieto, F. (2006) *PhD Thesis: Análisis de Sensibilidad y Optimización Aeroelástica de Puentes Colgantes en Entornos de Computación Distribuida.* University of La Corunna. (In Spanish).

[16] Nieto, F., Hernández, S. & Jurado, J.A. (2006) *Distributed Computing for Design Optimization of the Messina Bridge considering Aeroelastic Constraints.* The Fourth International Symposium on Computer Wind Engineering. Yokohame, Japan. July 16-19, 2006.

[17] Hernández, S., Hernández, L.A. & Antón, A. (2000) *Virtual Laboratories for Analysis and Design of Civil Engineering Structures.* Applications of High Performance Computing in Engineering VI. Wit Press.

[18] MAYA (2003) *Users Manual, 5.0 version.* Alias Wavefront.

Improvement analysis of long-span bridges flutter: Messina bridge example

J. Á. Jurado[1], A. León[2], F. Nieto[2] & S. Hernández[2]
[1]*School of Civil Engineering, University of Coruña, Elviña Campus, La Coruña, Spain*
[2]*University of Coruña, Elviña Campus, La Coruña, Spain*

Abstract

The hybrid methods used in the aeroelastic analysis of flutter for long-span bridges are computational, but they use coefficients and functions obtained experimentally in a wind tunnel. The experimental testings are carried out with a deck sectional model and there are two types: an aerodynamic testing for obtaining the aerodynamic coefficients such as drag, lift and moment, and an aeroelastic testing for obtaining the flutter derivatives, that the model oscillates under free vibration suspended by springs. In this paper, some improvements were achieved in the experimental phase as well as in the computational phase of the method. The influence of variation of the aerodynamic coefficients were studied with Reynold's number; different sets of springs were used to include a wide range of reduced velocities that the flutter derivatives depend on; the influence of deformation that the static wind load produces at the angle of attack along the bridge span was studied; finally this angle of attack was taken into account to determine the flutter derivatives used at each part of the deck. These improvements were applied to the sectional and computational models of the future Messina Strait Bridge in Italy.
Keywords: aeroelasticity, long-span bridges, flutter, sectional tests.

1 Introduction

The flutter condition on long-span bridges is critical during the design of these structures. To avoid experimental tests of completed bridge models in large wind tunnels that are complicated and expensive, it is necessary to use a hybrid method which is computational based but needs experimental parameters.

WIT Transactions on Engineering Sciences, Vol 58, © 2007 WIT Press
www.witpress.com, ISSN 1743-3533 (on-line)
doi:10.2495/EN070021

Sectional models of the deck are initially tested in an aerodynamic wind tunnel of smaller dimensions to obtain the flutter derivatives. These coefficients are then used in the computational analysis of the aeroelastic behaviour of the completed bridge.

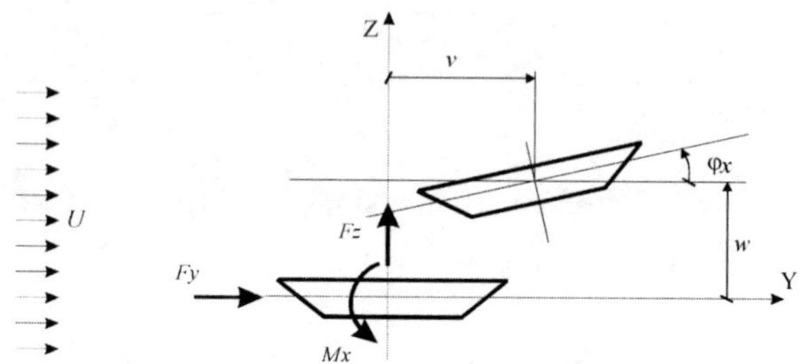

Figure 1: Forces and displacements of a sectional model.

Figure 1 shows the three forces acting on a deck. According to Simiu and Scanlan [1] formulation, these actions are linearized as functions of the displacements and velocities of the system for vertical w, lateral v and torsional rotation φ_x degrees of freedom. As it was explained in Jurado and Hernández (2000), the flutter condition is obtained by the computational solving of a non-linear eigen-problem which comes from the dynamic balance equation for the deck.

$$\mathbf{M}\ddot{\mathbf{u}} + \mathbf{C}\dot{\mathbf{u}} + \mathbf{K}\mathbf{u} = \mathbf{f}_a = \mathbf{K}_a\mathbf{u} + \mathbf{C}_a\dot{\mathbf{u}} \tag{1}$$

M, **C** and **K** are respectively the mass, damping and stiffness structural matrices. $\mathbf{f}_a$ is the aeroelastic forces vector which can be written assembling aeroelastic forces F_y F_z M_x along the deck as a stiffness $\mathbf{K}_a$ and damping $\mathbf{C}_a$ aeroelastic matrices multiplied by the displacements $\mathbf{u}$ and velocities $\dot{\mathbf{u}}$ vectors. The expressions of these matrices are

$$\mathbf{C}_a = \frac{1}{2}\rho U^2 KBl \cdot \begin{pmatrix} P_1^* & -P_5^* & -BP_2^* \\ -H_5^* & H_1^* & BH_2^* \\ -BA_5^* & BA_1^* & B^2A_2^* \end{pmatrix} \quad \mathbf{K}_a = \frac{1}{2}\rho U^2 K^2 l \cdot \begin{pmatrix} P_4^* & -P_6^* & -BP_3^* \\ -H_6^* & H_4^* & BH_3^* \\ -BA_6^* & BA_4^* & B^2A_3^* \end{pmatrix} \tag{2}$$

where B is the deck width, ρ is the air density, U is the mean wind speed, $K = B\omega/U$ is the reduced frequency with ω the frequency of the response and $H^*_i(K)$, $P^*_i(K)$, $A^*_i(K)$ i = 1...6 are the flutter derivatives which are functions of K.

In the following sections, the improvements made in the sectional bridge deck testings are explained as well as the improvements in the computational calculation of the critical flutter velocity for the entire bridge structure. Those advances are applied to the future Messina strait bridge model, between Sicily and the Italian peninsula.

2 Aerodynamic sectional testing of the Messina Bridge

In order to carry out the static analysis of the wind load, we need to obtain previously the aerodynamic coefficients of the deck cross section in function of the angle of attack. The aerodynamic coefficients are obtained by carrying out a testing of a fixed deck sectional model measuring inside the wind tunnel drag force D_s, lift L_s and moment M_s that exerts air flow over the model. See figure 2.

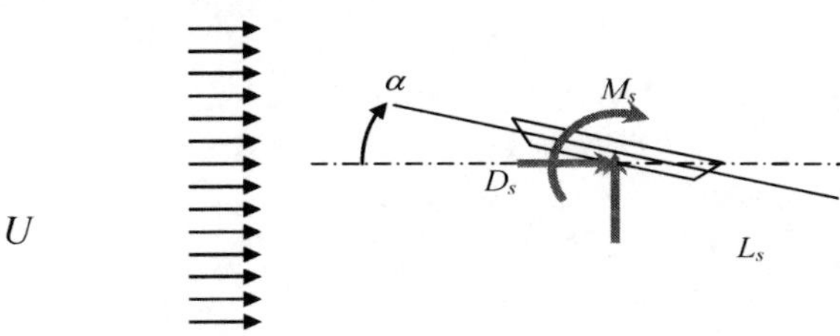

Figure 2: Aerodynamic forces on a sectional model.

First of all, a scale bridge deck sectional model is built, whose shape should be as similar to the prototype as possible. Figure 3 shows a sectional model of the Messina Bridge. The mass does not need to be scaled to the original bridge since the displacements are constrained and there are no inertial forces. We simply search for simulating the real boundary conditions that determine the air flow around the deck.

Figure 3: Sectional deck model of the Messina Strait Bridge.

An important problem for obtaining the aerodynamic coefficients is its dependence on Reynolds number (Re = $\rho UB/\mu$.; ρ: air density, m: air viscosity; U: wind velocity, B: deck width), see Son and Hanratty [2]. As a consequence the aerodynamic forces and the aerodynamic coefficients change noticeably when the wind velocity increases as shown in Figure 2. From a certain wind velocity value, the coefficient values are quite stable. It is essential to determine the correct air flow velocity in the wind tunnel to obtain the aerodynamic forces. This should be carried out by trying with different angles of attack as shown in Figure 4 for the Messina Strait Bridge. In the graph, it can be observed that with Reynold's number over 400000, the values do not vary noticeably. Once the velocity is determined, 11m/s for the Messina example which corresponds to Re = 460000, the testings were carried out varying the angle of attack to obtain the aerodynamic coefficient graphs like the one shown in figure 5.

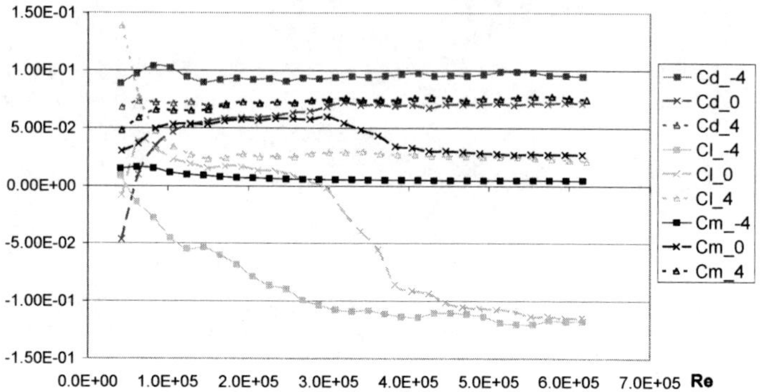

Figure 4: Variation of the aerodynamic coefficients of the Messina Bridge deck for different angles of attack in function of Reynold's number.

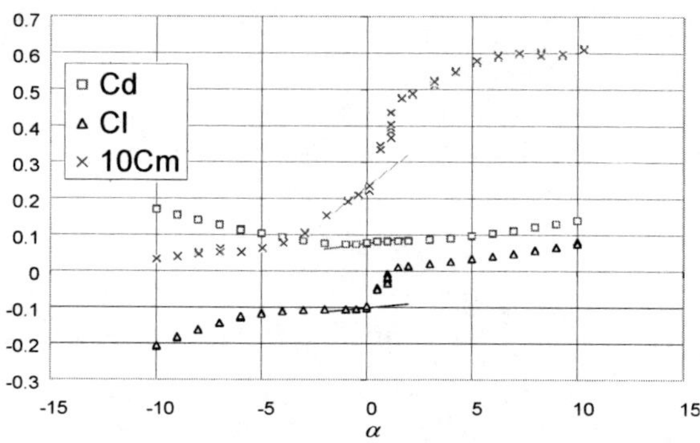

Figure 5: Aerodynamic coefficients of the Messina Bridge deck.

3 Aeroelastic sectional testing of the Messina Bridge

The aeroelastic sectional testings under free vibration consist of sustaining the sectional model by springs and make them freely oscillate with and without air flow inside the wind tunnel. A detailed explanation of this procedure is found in Jurado et al [3]. From the displacements of the model, we can calculate the stiffness properties and damping, for example with the MITD (Modified Ibrahim and Mikulcik [4] Time Domain Method), and from the variation of those properties as the wind velocity varies, we can obtain the flutter functions. See Sarkar et al [5].

A sectional model requires less similarity conditions than a reduced complete bridge model. The most important thing is to maintain the geometric similarity. It is recommended that the length of the model is three times its width in order to

be considered two-dimensional. Scales less than 1/100 should not be used. The mass and inertia values are not very important for the testing even though it is recommendable that they are small enough for not entering errors in measuring the wind forces. Therefore there is no need to consider a scale of masses since we only try to quantify the wind action in function of the oscillatory movements of the deck. The model is elastically sustained using eight to twelve springs: four or eight vertical and four horizontal ones (Figure 6).

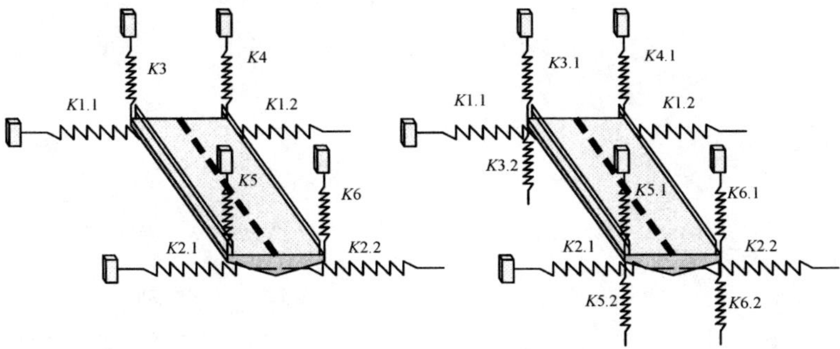

Figure 6: Sustentation of the sectional model with eight or twelve springs.

The stiffness of the springs determines the vibration frequencies ($2\pi f = \omega$) of the system that together with the wind velocity in the tunnel, V and the model width, B, determines the range of reduced velocities, V^*, in order to be able to obtain flutter functions.

$$V^* = V/fB = 2\pi V/\omega B = 2\pi/K \quad (3)$$

For example, eighteen flutter functions of the deck model of the Messina strait bridge were obtained considering three aerodynamic appendages at the wind shields. Testings were carried out varying the angle of attack between -3°, 0° and +3°. Four types of sustaining the model with different springs were used in order to vary the natural frequencies of the model and include a wider range of reduced velocities. For the first three testings, the model was allowed to have vertical and horizontal displacements and rotation, while in the fourth one, it was only allowed to rotate. The model for the first testing was sustained with twelve springs, and for the second and the third testings it was sustained with eight springs. For the fourth testing, the model was not allowed to move vertically or horizontally with two bars in each side of the model. The mass of the model is 7.47 kg and its torsional moment of inertia is 0.35 kg·m^2, and the natural frequencies of the system for each sustaining type are the followings.

The testings were carried out for wind velocities between 6 and 20 m/s. The minimum and maximum reduced velocities with which flutter functions can be obtained are shown in Table 2. With the testings of three degrees of freedom, 18 flutter functions are obtained simultaneously, while with the testing of only rotational freedom, we can only identify the flutter functions, A_2^* and A_3^*. Figure 7 shows the obtained flutter functions.

Table 1: Natural frequencies of the aeroelastic testings.

	fv(Hz)	fw(Hz)	$f\varphi_x$(Hz)
Test 1	2.2	2.9	6.8
Test 2	2.8	1.7	3.5
Test 3	1.3	1.4	2.0
Test 4	0	0	1.14

Table 2: Range of reduced velocities for obtaining different flutter functions.

	(A_5*, A_6*, H_5*, H_6*, P_1*, P_4*)		(A_1*, A_4*, H_1*, H_4*, P_5*, P_6*)		(A_2*, A_3*, H_2*, H_3*,P_2*, P_3*)	
	$V^*(f_v)$ min	$V^*(f_v)$ max	$V^*(f_w)$ min	$V^*(f_w)$ max	$V^*(f_{\varphi x})$	$V^*(f_{\varphi x})$
Test 1	4.521	15.07	3.40	11.34	1.45	4.85
Test 2	3.54	11.79	5.82	19.41	2.82	9.39
Test 3	7.62	25.38	7.22	24.09	4.90	16.34
Test 4	-	-	-	-	8.68	28.95

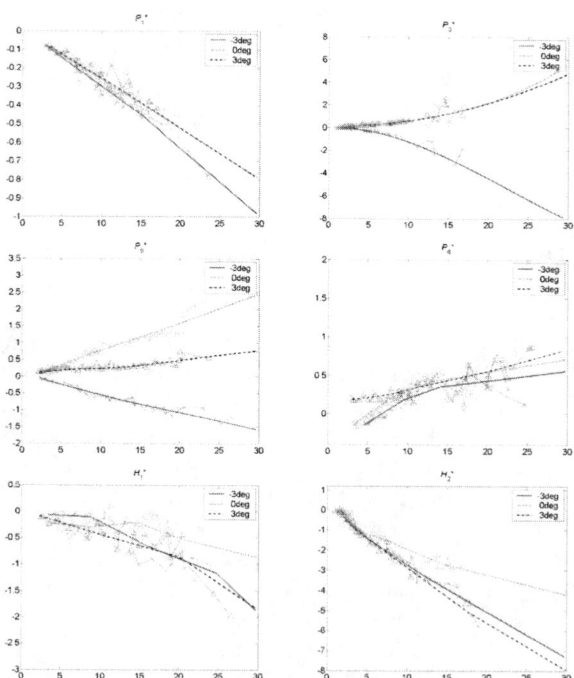

Figure 7: Same flutter derivatives of the deck of the Messina bridge in function of the reduced velocity $V^* = 2\,\pi / K$ (x axis).

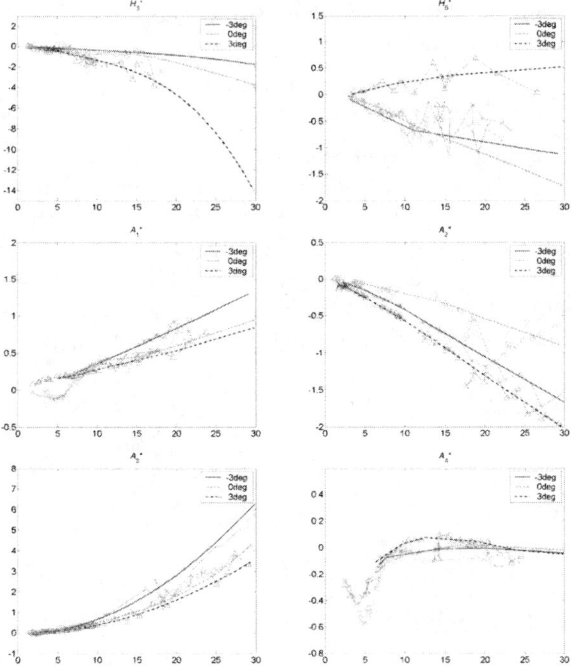

Figure 7: Continued.

4 Influence of the static deformation on the angle of attack

A non-linearity that normally not taken into account for wind load analyses of long-span bridges is the variation of angle of attack due to deck deformation. Rotation about the axis of the deck can vary some degrees along the deck; for example, 0.6° for the Messina Bridge [6] at the centre span, and 3° for the Akashi Bridge. The deck rotation causes a change in the angles of attack, which affects the aerodynamic coefficient and the flutter coefficient values. Calculation of the static solution of the wind is carried out by resolving the equation system that relates the forces on the structure and the displacements. This system can be expressed in matrix form as:

$$\mathbf{K}(\mathbf{u})\cdot\mathbf{u} = \mathbf{f}\left[\alpha(\mathbf{u})\right] \tag{4}$$

where not only the stiffness matrix, **K** depends on the displacements, **u**, but also the wind loads, **f** that vary with the angle of attack, α throughout the deck and therefore also depend on **u**. If this dependency is not taken into account, (4) is converted in a linear problem in which the stiffness matrix and the wind force are used for the initial positions of the nodes with null angle of attack.

An approximation of the problem (4) consists of considering the deck deformation in the wind forces.

$$\mathbf{K}\cdot\mathbf{u} = \mathbf{f}\left[\alpha(\mathbf{u})\right] \tag{5}$$

In this approximation, the influence of the wind forces on the structural stiffness is not considered. The problem is resolved by iterations. In each step, new forces that depend on the obtained displacements in the previous step are defined. The process finishes when the obtained displacements converge, that is to say they do not change too much between two iterations. This method was applied to the static deformation of the Messina Bridge observing noticeable differences in the deck displacements for the wind velocity over 50m/s as shown in Figure 8.

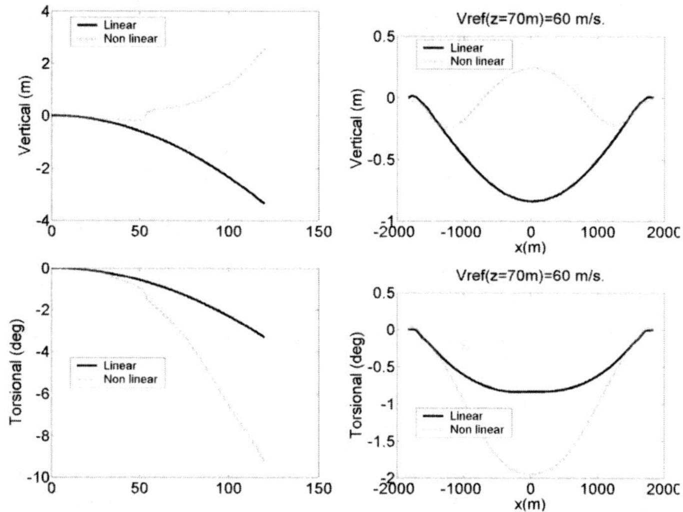

Figure 8: Deck displacements of the Messina Bridge at increasing wind velocity.

5 Flutter analysis of Messina Bridge

In the present work, a coherent matrix formulation has been used for hybrid flutter analysis. Jurado and Hernandez [7] explain this formulation stems from the equation (1). Through modal analysis it is possible to approximate the deck displacements by means of a linear combination of the most significant mode shapes. Assembling them in columns into the modal matrix $\mathbf{\Phi}$, the displacement vector can be expressed as $\mathbf{u} = \mathbf{\Phi q}$. Each element of the vector $\mathbf{q}$ represents the participation of each mode shape in the displacement vector $\mathbf{u}$. Premultiplying (1) by $\mathbf{\Phi}^T$ it becomes

$$\mathbf{I}\ddot{\mathbf{q}} + \mathbf{C}_R\dot{\mathbf{q}} + \mathbf{K}_R\mathbf{q} = \mathbf{0} \tag{6}$$

where $\mathbf{C}_R = \mathbf{\Phi}^T (\mathbf{C} - \mathbf{C}_a) \mathbf{\Phi}$, $\mathbf{K}_R = \mathbf{\Phi}^T (\mathbf{K} - \mathbf{K}_a) \tilde{\mathbf{\Phi}}$ and $\mathbf{\Phi}^T \mathbf{M} \mathbf{\Phi} = \mathbf{I}$ Using mass normalized modes. Knowing that the solution of this equation has the form $\mathbf{q}(t) = \mathbf{w}e^{\mu t}$, becomes

$$\left(\mu^2 \mathbf{Iw} + \mu \mathbf{C}_R \mathbf{w} + \mathbf{K}_R \mathbf{w}\right) e^{\mu t} = \mathbf{0} \tag{7}$$

which can be transformed into an eigenvalue problem by adding the identity $-\mu\mathbf{Iw}+\mu\mathbf{Iw} = \mathbf{0}$:

$$\left[\mu \begin{pmatrix} \mathbf{I} & \mathbf{0} \\ \mathbf{0} & \mathbf{I} \end{pmatrix} \begin{pmatrix} \mu\mathbf{w} \\ \mathbf{w} \end{pmatrix} + \begin{pmatrix} \mathbf{C}_R & \mathbf{K}_R \\ -\mathbf{I} & \mathbf{0} \end{pmatrix} \begin{pmatrix} \mu\mathbf{w} \\ \mathbf{w} \end{pmatrix}\right] e^{\mu t} = \mathbf{0} \tag{8}$$

or in short

$$\left(\mathbf{A} - \mu\mathbf{I}\right) \mathbf{w}_\mu e^{\mu t} = \mathbf{0} \tag{9}$$

The imaginary part of the eigenvalues μ counts on the frequency ω, while the real part of the eigenvalues is associated with the damping ratio ξ. The condition of flutter corresponds to the lowest wind speed U_f which gives one eigenvalue with vanished real part. However, the problem (9) is non-linear because the matrix $\mathbf{A}$ assembles the aeroelastic matrices $\mathbf{K}_a$ and $\mathbf{C}_a$. These matrices contain the flutter derivatives, which are functions of the reduced frequency $K = B\omega/\bar{U}$, and the frequency for each eigenvalue ω remains unknown until the problem has been solved.

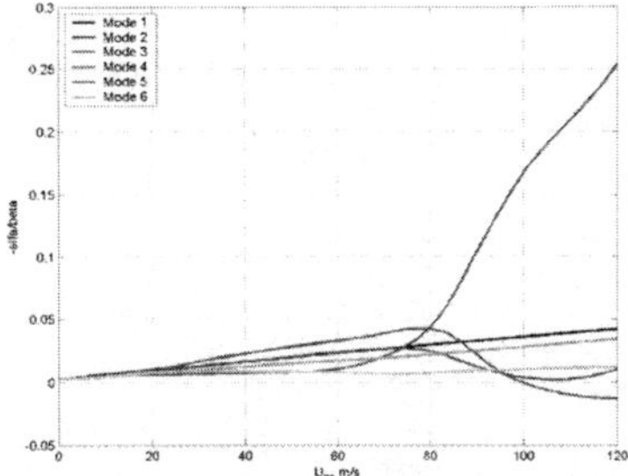

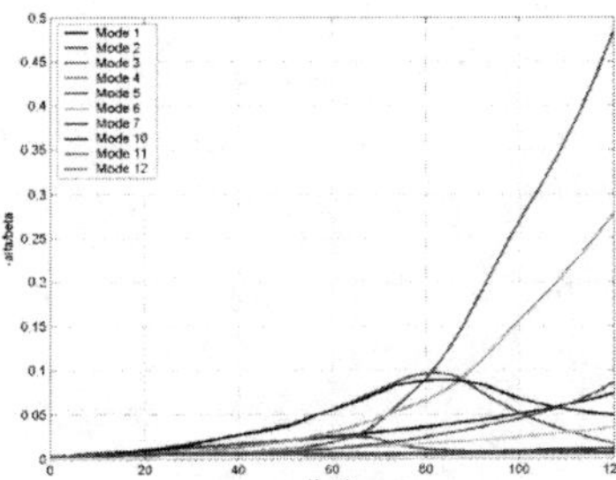

Figure 9: Relation between the real and imaginary part of the eigenvalues in the flutter analysis not considering (left U_f = 100m/s) and considering (right U_f> 120m/s) the variation of angle of attack.

An improvement of the bridge flutter analysis consists of taking into account the dependency of the flutter derivatives on the angle of attack at each point of the deck due to the static deformation, and its calculation is explained in the previous section. Figure 9 shows graphs with the evolution of the proper values of the problem (9) for increasing wind velocities with or without varying the angle of attack. It is observed that the critical flutter velocity increases favourably when this effect is considered.

6 Conclusions

The wind velocity used for the aerodynamic sectional testing in a wind tunnel should be chosen by previously studying the variation of aerodynamic forces in unction of Reynold's number.

If various sets of springs are used in the aeroelastic sectional testings, a bigger range of reduced velocities can be included in obtaining the flutter derivatives. The static deformation due to the wind load affects considerably the angle of attack at high wind velocities close to the flutter.

The critical flutter velocity also varies noticeably if the variation of angle of attack along the bridge span is taken into account. The flutter derivatives should be chosen according to this dependency.

Acknowledgement

This research has been funded by the Galician Council of Innovation, Trade and Industry under project PGIDIT04CCP118002PR.

References

[1] Simiu E., Scanlan R. H., (1996) *Wind Effect on Structures*, Wiley N. Y.

[2] Son J. S., Hanratty T. J., (1969) *Numerical solution for the flow around a cylinder at Reynolds number of 40, 200, 500.* J. of Fluid Mech. Vol. 35 pp: 369-386.

[3] Jurado J. Á., León A., Hernández S. (2005) *Wind Tunnel Control Software for Identification of Flutter Derivatives on Bridge Sectional Tests.* EACWE 4. 4° European and African Congress in Wind Eng. Prague, Check Republic.

[4] Ibrahim S.R., Mikulcik E. C. (1977) *A Method for the Direct Identification of Vibration Parameters from the Free Response*, The Shock and Vibration Bulletin, bulletin 47, Part 4.

[5] Sarkar P. P., Jones N. P., Scanlan R. H. (1992) *System identification for estimation of flutter derivatives,* Journal of Wind Engineering and Industrial Aerodynamics 41-44 1243-1254.

[6] Stretto di Messina S.p.A. (2004), Specifiche tecniche per il progetto definitivo e il progetto esecutivo dell'opera di attraversametnto. Requisiti e linee guida per lo sviluppo della progettazione. GCG.F.05.03.

[7] Jurado, J. Á., Hernandez S. (2004) *Sensitivity analysis of bridge flutter with respect to mechanical parameters of the deck.* Structural and Multidisciplinary Optimization Vol. 27, N° 4. June, 2004.

Seismic analysis: past, present and the future

D. Mestrovic, D. Cizmar & P. Roncevic
Faculty of Civil Engineering, University of Zagreb, Croatia

Abstract

This paper summarizes earthquake calculations according to Eurocode regulations and codes applied in Croatia (Europe). The paper shows past, present and future methods for determination of seismic forces – an overview of all methods used is given. A simple pseudo static calculation is analyzed and compared to seismic forces according to Eurocode. An overview of spectral analysis according to Eurocode and Croatian code is given and discussed. Finally, time history analysis is discussed because as computer power increases with the availability of the strong motion database, the real records are becoming more popular for defining the input to dynamic analyses. The paper discuses three basic methods of input data – artificial records, synthetics accelograms and real records recorded during earthquakes. The possibility of using real records (for both linear and non-linear time history) is analyzed and discussed. A conclusion about the use of real accelograms (time history) versus spectral analysis is given.
Keywords: seismic analysis, time history, accelogram.

1 Introduction

An earthquake is a phenomenon that results from the sudden release of stored energy in the Earth's crust that creates seismic waves. At the Earth's surface, earthquakes manifest themselves by a shaking or displacement of the ground. Earthquakes are related to the tectonic nature of the Earth. The Earth's lithosphere is a composed of plates in very slow but constant motion caused by the heat in the Earth's mantle and core. Plate boundaries can grind past each other, creating frictional stress. When the frictional stress exceeds a critical value, a sudden failure occurs. The boundary of tectonic plates along which failure occurs is called the fault plane. When the failure at the fault plane results in a displacement of the Earth's crust, the elastic strain energy is released and

WIT Transactions on Engineering Sciences, Vol 58, © 2007 WIT Press
www.witpress.com, ISSN 1743-3533 (on-line)
doi:10.2495/EN070031

seismic waves are radiated, causing an earthquake. The severity of an earthquake is described by magnitude and intensity. Magnitude characterizes the size of an earthquake by measuring indirectly the energy released (Richter scale). Intensity indicates the local effects and potential for damage produced by an earthquake on the Earth's surface as it affects humans, animals, structures and natural objects. To be able to show the extent of various levels of seismic effects within a particular location, seismologists usually compile maps called isoseismic maps. An isoseismic map uses contours to outline areas of equal value in terms of ground shaking intensity, ground surface liquefaction, shaking amplification, or other seismic effects. Fig. 1 shows a typical isoseismic map. On the other hand, in building codes, the isoseismic maps are converted into seismic zone maps, which are used for seismic analysis of structural components of buildings. The seismic zone maps show the severity of expected earthquake shaking for a particular level of probability.

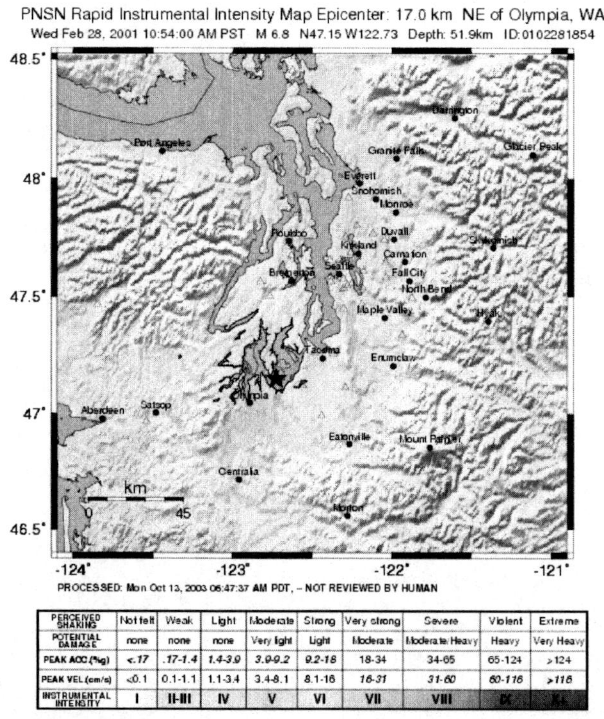

PERCEIVED SHAKING	Not felt	Weak	Light	Moderate	Strong	Very strong	Severe	Violent	Extreme
POTENTIAL DAMAGE	none	none	none	Very light	Light	Moderate	Moderate/Heavy	Heavy	Very Heavy
PEAK ACC.(%g)	<.17	.17-1.4	1.4-3.9	3.9-9.2	9.2-18	18-34	34-65	65-124	>124
PEAK VEL.(cm/s)	<0.1	0.1-1.1	1.1-3.4	3.4-8.1	8.1-16	16-31	31-60	60-116	>116
INSTRUMENTAL INTENSITY	I	II-III	IV	V	VI	VII	VIII	IX	X+

Figure 1: An isoseismic map (source: United States Geographical Survey).

2 Model of structure

This paper analyzes a simple five-storey reinforced concrete building which is situated in a seismically active region of Croatia. The structure is modeled as a simple plane frame. Each frame is made of concrete C25 according to

Eurocode 2. Columns are 60/60 cm for the first two stories, other columns (stories three to five) are 40/40 cm. Beams are for the first storey 30/70 cm, for all others 55/70 cm. Fig. 2 shows a model of the structure (units are in centimeters).

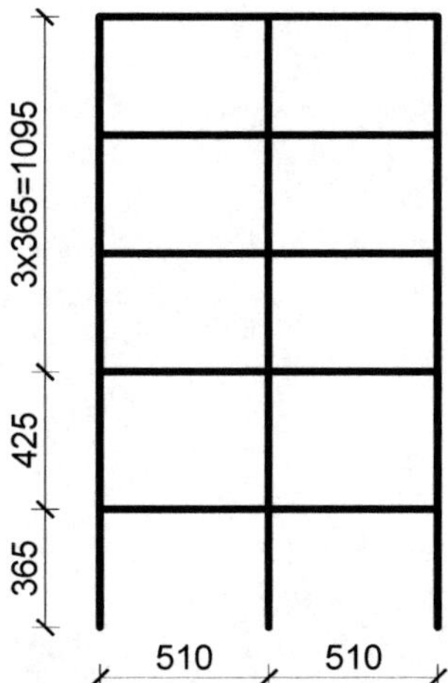

Figure 2: Model of concrete frame.

3 Seismic analysis – overview of methods

The focus of this paper is the comparison between different methods of obtaining seismic forces. The paper will analyze a five-storey reinforced concrete building which is situated in a seismically active region of Croatia. The structure is modelled as a simple plane frame. Several different calculations are made (pseudo static, spectrum analysis, non-linear time history and pushover). Pseudo-static analysis is made according to Croatian regulations. The structure is also analyzed using spectrum analysis according to EC8 (for this type of analysis both the elastic and design spectra are computed and discussed). Non-linear time history analysis is made using two synthetic accelograms that represent earthquakes in this region. Based on available digitized accelograms, two earthquakes were selected according to their magnitudes: an earthquake occurring in Petrovac (Montenegro) on April 15, 1979, with magnitude M=6.8 and an earthquake occurring on April 9, 1979, in Ulcinj (Montenegro) with magnitude M=5.3.

3.1 Pseudo-static analysis

Pseudo-static analysis is made according to Croatian regulations. Total base shear (S) is calculated according to Eqn. (1) (K is the total seismic coefficient, G is the total weight of structure, G_i is the weight of each floor). The total seismic coefficient is given in Eqn. (2) (K_o is the coefficient depending on the class of

the structure, K_s depends on the seismic intensity, K_d depends upon the natural period of the structure and K_p is the coefficient that accounts for ductility and damping). For the purpose of this analysis K_S=0.1 and all other coefficients are equal to 1, so K=0.1.

$$S=K\cdot G \tag{1}$$

$$K= K_o\cdot K_S\cdot K_d\cdot K_p \tag{2}$$

For structures up to five stories the seismic force for each floor (S_i) is calculated as in Eqn. (3). For structures with more than five stories 15% of the total force is applied at the top level and the remaining 85% is calculated as in Eqn. (3).

$$S_i = S\cdot\frac{G_i\cdot H_i}{\sum G_i\cdot H_i} \tag{3}$$

3.2 Spectrum analysis

Frames are also analyzed using spectrum analysis according to EC8. The ground type is A (rock or other rock-like geological formation). Damping is assumed to be 5%. Fig. 4. shows the spectrum functions that were used as input for spectrum analysis (both the elastic and design spectra are computed). For the design spectrum a behavior factor of 4 is assumed.

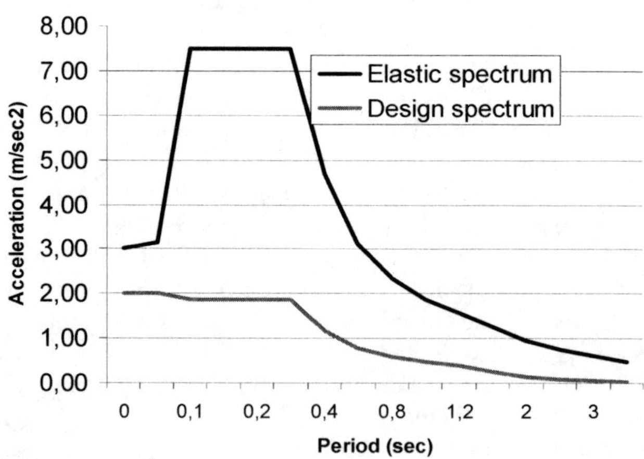

Figure 3: Spectrum function used for spectral analysis.

3.3 Linear and non-linear time history analysis

There are three basic options available for obtaining acceleration time-series: artificial earthquakes generated by software, synthetic accelograms and real records from earthquakes. For artificial earthquake generation a common approach is to generate a power spectral density function from the smoothed response spectrum, and then to derive sinusoidal signals having random phase angles and amplitudes. In this way we can generate many earthquakes very easily. But we should never forget that there is a big problem with artificial records: they usually have an excessive number of cycles of strong motion and they possess high energy content. Secondly, there is a possibility to generate a synthetic accelogram. Synthetic accelograms are generated from seismological source models and account for path and site effects. Models for generation of synthetic accelograms are very complex and usually an engineer must work with the seismologist. The third and most interesting to us are real records recorded during earthquakes. These accelograms are free from all problems associated with artificial spectrum-compatible records. They are better than synthetic accelograms because little knowledge of seismology is required. Real records are now easily accessible in large numbers, but for a specific location there aren't many recorded earthquakes.

For this paper linear and non-linear time history analyses are made using a synthetic accelogram that represents an earthquake in this region. Based on available digitized accelograms, two earthquakes were selected according to their magnitudes: an earthquake occurring in Petrovac (Montenegro) on April 15, 1979, with magnitude M=6.8 and an earthquake occurring on April 9, 1979, in Ulcinj (Montenegro) with magnitude M=5.3. From these earthquakes, the synthetic accelogram was generated. A calculation was made for a synthetic time sequence of accelerations which, on the average, correspond to earthquakes with magnitude 7.0 and epicentre distance 0 km. 10 m hypocenter depth was generally assumed. Fig. 4 represents an accelogram which was used for both linear and non-linear analyses.

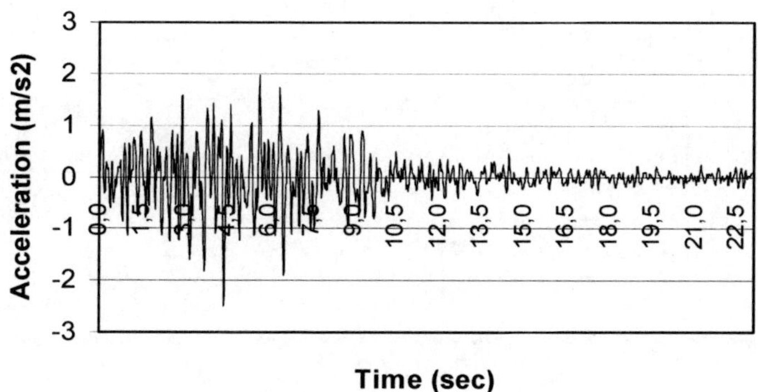

Figure 4: Accelogram used for time history analysis.

3.4 Pushover analysis

Non-linear pushover analysis is made for seismic assessment of the structure. The analysis was displacement controlled. The maximum displacement was 70 cm. The non-linear behavior occurs in discrete hinges. Hinges are introduced

into the frame and assigned at the end of columns and beams. A coupled hinge that yields results based on the interaction of the axial force and the bending moments at the hinge location is used.

Table 1: Results of different types of analyses.

Type of analysis	Total base shear (kN)	Maximal column shear (kN)
Pseudo static	181.6	79.7
Spectrum - elastic	192	70.0
Spectrum - design	52.7	21.7
Linear time-history	603.9	248.5
Non-linear time history	445.4	173.6

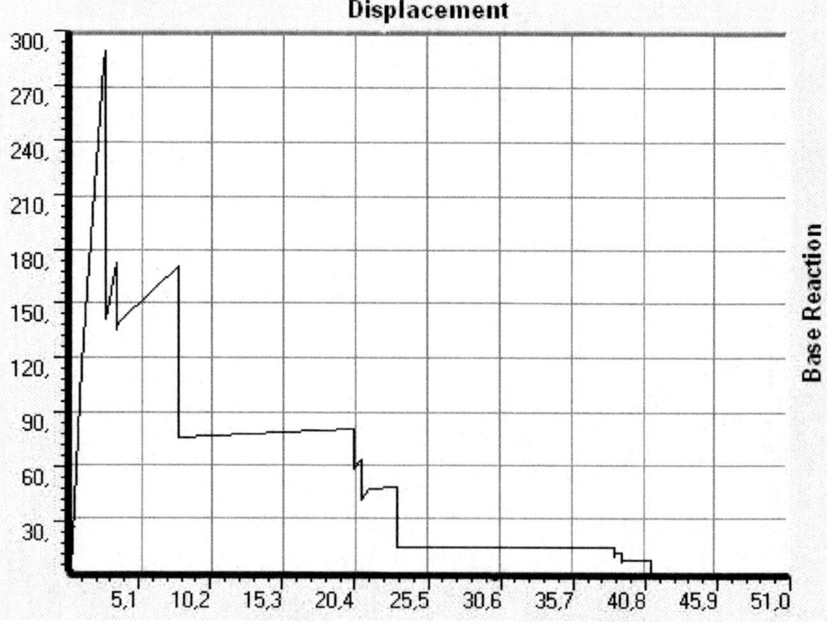

Figure 5: Pushover curve.

4 Results

Table 1 show results for a five-storey frame. Different types of analyses show very large numerical differences. It is evident that the design spectrum has smallest seismic forces (base shear is around 3% of total vertical load). This is because of the relatively great behaviour factor that was assumed. On the other hand elastic spectrum analysis has a base shear around 12.6%. Pseudo static analysis has 10% base shear. Due to the very strong earthquake generated, the linear time history has the greatest seismic forces. Non-linear time history analysis has around 40-50% less forces, mainly due to the non-linear distribution of forces. Fig. 5 shows the pushover curve (base reaction vs. displacement) for the same frame. It is very clear how the seismic capacity of the structure is decreasing with the increase of the horizontal displacement.

5 Conclusion about the future of seismic analysis

Today, the use of pseudo-static analysis is totally abandoned. Most structures are calculated using spectrum analysis, which is very easy to perform, and time history. Time history and pushover analysis are used more often for highly irregular structures and structures for which higher modes are likely to be excited. So now many questions arise: should we abandon spectrum analysis and use only time history and pushover? The answer is yes, but only when sufficient data is available. When using time history, we suggest using several accelograms with different epicenter distances. With more real earthquakes in a database it will be possible to conduct real accelograms and full time history. Other ways to perform full dynamic analysis is to generate synthetic accelograms for each seismic region.

References

[1] J. J. Bommer, A.B. Acevedo, "The use of real accelograms as input for dynamic analysis", Journal of Earthquake Engineering Vol. 8, London, 2004.

[2] W. Spence, S. A. Sipkin, G. L. Choy, “Measuring the size of earthquake”, Unites States Geographical Survey, Reston, 1989.

[3] D. Cizmar, A. Nizic, D. Mestrovic, “Importance of dynamic characteristics of accelograms to structural response”, SECED Conference, 2005.

[4] A. Mihanovic, “Dynamics of structures”, University of Split – Faculty of Civil Engineering, 1995.

[5] SeismoSoft [2004] "SeismoSignal - A computer program for signal processing of strong-motion data" [online]. Available from URL: http://www.seismosoft.com.

[6] M. Causevic, “Earthquake engineering”, University of Rijeka, Rijeka, 2001.

[7] E. Prelogovic, B. Aljinovich, and S. Bahun, “New data on structural relationship in the north Dalmatian Dinaride area”: Geologia Croatica, v. 48, no. 2, 1985

Seismic analyses of the Messina Bridge project

S. Hernández, L. E. Romera, A. Baldomir & F. Bravo
School of Civil Engineering, University of Coruña, Spain

Abstract

In this paper several preliminary works developed to determine the seismic response of the Messina Bridge project are presented. Firstly the different types of non-linear structural models developed are shown, including global models for the macro-level static and dynamic analysis, and local models for the micro-level analysis. Then, the analyses and several results obtained for static and nonlinear seismic time histories analyses are explained. Finally, the results of the Messina local tower model with a longitudinal push-over analysis are presented.

1 Introduction

The suspension bridge over the strait of Messina (figure 1) will link Sicily and Calabria with the world longest center span of 3.300 m. The preliminary design used as the basis of the analysis that appear in this article, was developed by Chodai Co., Ltd. in 2004, following the analysis and design specifications for the bridge (ref.TSA002) developed by the concessionaire Stretto Di Messina S.p.A which carries out the research, design, construction and operation of the bridge.

Figure 2 shows the transversal bridge section and one of the towers. The bridge section has three principal girders, one central for two railways and two laterals for road traffic. These girders are connected with transversal girders spaced 30 m. in the longitudinal direction, and suspended at its ends from hangers that connect them with the main cables. The main cables are connected to the ground at the splay saddle in the anchorage, and to the tower saddle at the towers. Figure 3 shows a realistic visualization detail of these girders.

The towers as well as the girders and the suspension system were executed in structural steel of qualities between S355J0 and S420J0. Each tower has an approximate height of 382 m, with a self-weight greater than 5 x 10^5 kN, and is composed of two inclined columns connected by four transversal girders.

WIT Transactions on Engineering Sciences, Vol 58, © 2007 WIT Press
www.witpress.com, ISSN 1743-3533 (on-line)
doi:10.2495/EN070041

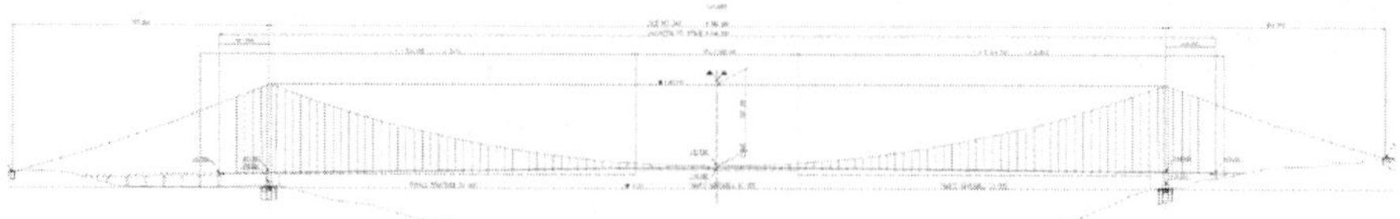

Figure 1: Front and top view of Messina Bridge.

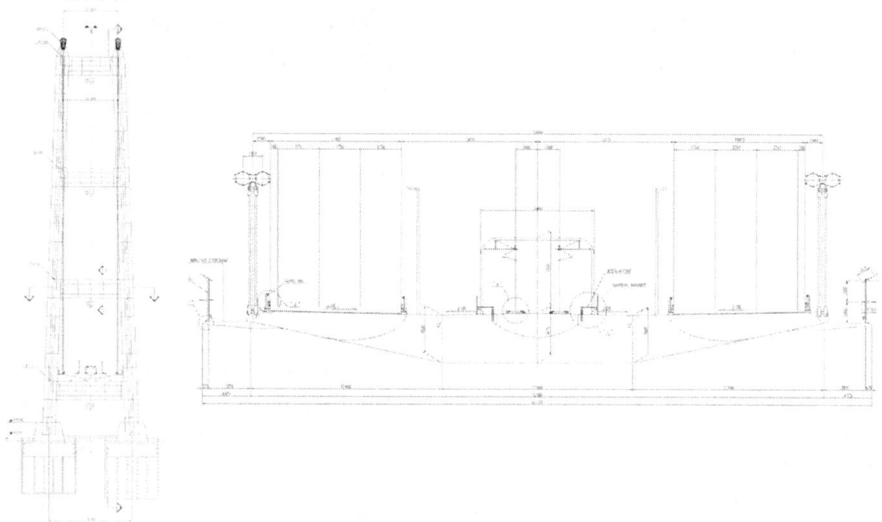

Figure 2: Messina tower and bridge central cross-section.

Figure 3: Visualization model: top and bottom aerial views.

The loads considered in design specifications are:

- Dead loads: permanent loads (PP) and non-structural loads (PN)
- Live loads: variable man-generated actions (Q) due to road and railway loads, for local resistance and deformations checks at micro-level (QL), at macro-level (QA), and for runnability (QR); and variable environmental actions V due to wind action (VV), seismic action (VS) and thermal actions (VT)

- Imperfections: fundamentally levels of verticality and rectilinearity errors for the shafts of the towers
- Accidental loads (A)

These loads and their combinations are specified in four limit states:

- SLS1: Limit service state with road and railway runnability and no damage.
- SLS2: Limit service state with only railway runnability and minimal damage.
- SLU: Ultimate limit state with lack of functionality in road and railway, and reparable damage.
- SLIS: Limit state of structural integrity with significant damage.

Due to the great dimensions of the bridge, the static and aero-elastic wind loads are some of the fundamental conditions in the design of the bridge, determining the principal girder geometry. On the other hand the bridge location is one of the most important areas of seismic activity in Europe. The Messina earthquake in 1908 was the deadliest quake in European history, with a magnitude equalled to 7.5 in Ritcher scale and with one return period characterized between 1000/1500 years, caused more than 100000 deaths. The earthquake and the later tsunami destroyed the city of Messina. Therefore, seismic loads are very high and they determine the design of the bridge towers besides the wind loads.

2 Structural models

To carry out the static and dynamic analysis one global model of the complete bridge and several local models (of parts) of the bridge were developed, using the commercial finite element programs Cosmos/m v.2.9 [1] and Abaqus v.6 [2].

Figure 4 shows the global model developed using 3-dimensional beam and truss elements to model longitudinal and transversal girders, towers, hangers, main cables and main cable anchorages; linear springs in connections at base tower representing foundation and soil stiffness, and several coupling conditions between degrees of freedom of the girders and the connection system to the towers are applied to ensure the correct behaviour of the connection system. The model has 2848 nodes and 3848 elements and represents initially the bridge under full dead load, with initial stresses due to this load in the main suspension cables, hangers and towers, and the design geometrical configuration, with null shear force at the tower tops.

The global model allows us to develop static and dynamic nonlinear analysis considering the fully nonlinear geometrical behaviour [3, 4] due to the geometrical stiffness, the linear stiffness and the contributions by changes in the geometrical configuration. The first step is always an equilibrium iteration check between the complete set of dead loads and the initial stresses, giving a result of zero displacements. To calculate the distribution of initial stresses in equilibrium with dead loads, an iterative program was developed to determine the initial length and geometrical configuration of the main suspension cables with null initial stress, which gives the design configuration under dead loads.

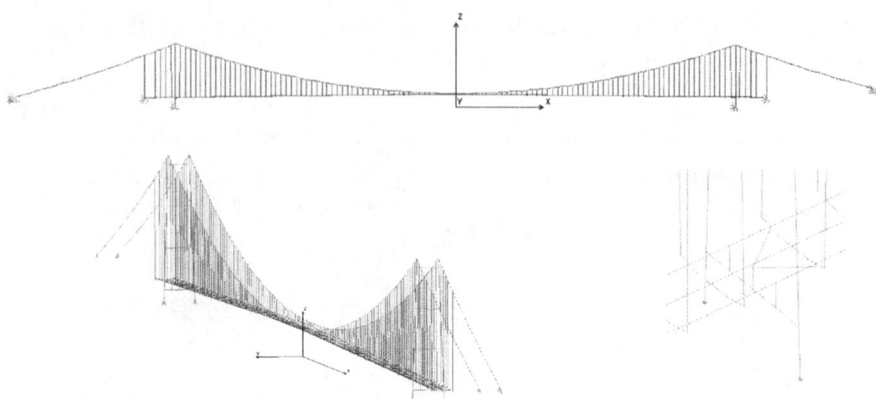

Figure 4: Views of global model and detail of connections between girders and towers.

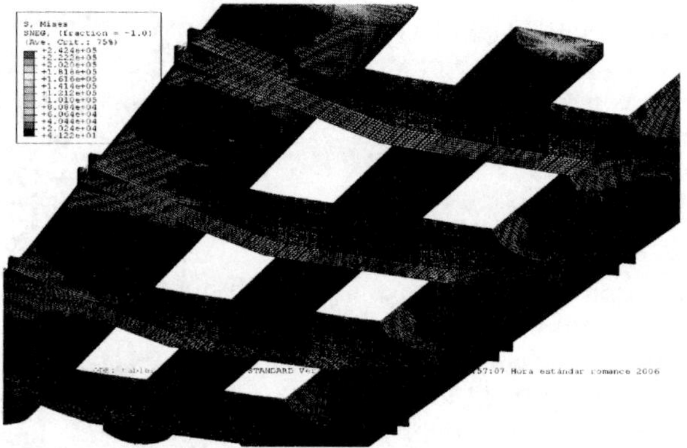

Figure 5: Local model of longitudinal girders and four transverse girders.

Due to the great dimensions of the bridge, the usual method of analysis in two steps: an initial linear step with dead loads and a second analysis including the effect of the geometric stiffness matrix calculated with the stress results obtained in the first step, produces a wrong vertical displacement in the middle of the span greater than 32 m.

Subsequent nonlinear steps allow us to introduce static or seismic loads. The results of displacements, forces and stresses are used for global checks or as resultant global forces applied in local models to study in detail the behaviour of several parts including geometrical and material nonlinearity. Figure 5 shows a local model of girders section made of more than 500.000 shells elements, including all the internal stiffeners as shell elements. In the limits of the section, forces and moments obtained from the global model in the corresponding beam nodes are introduced at the center of gravity of girders by kinematics coupling conditions, and were distributed over the section girder nodes. In the same manner, hanger forces and loads that affect to the modelled shell elements are considered to achieve the equilibrium in a fully nonlinear analysis, allowing us to study in detail the plastic behaviour and local buckling effects.

In the case of the towers, the Messina tower was first modelled with a local model (figure 6a) using 3D beam elements, and springs representing the foundation stiffness in the bottom nodes and springs in the top connection with the main cables calculated from the global model (figure 6b) applying several levels of horizontal, transversal and vertical loads at the top of the tower.

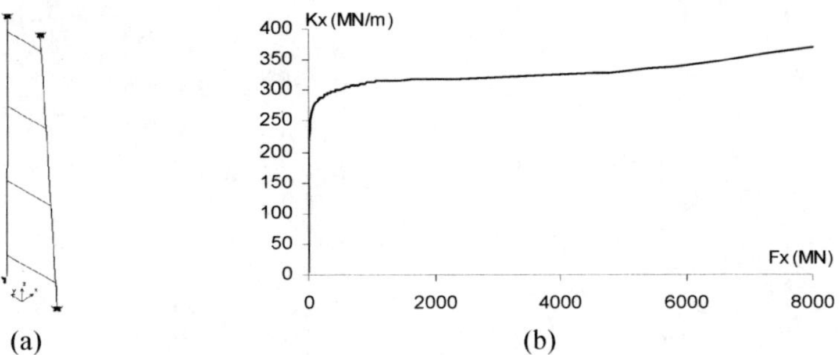

Figure 6: Local beam model of the Messina tower (a); and longitudinal behaviour for the top springs that model the connection with the suspension cables (b).

Then a complex shell model of half tower, with more than 80.000 shell elements (figure 7), was developed to analyze the plastic behaviour, possible local buckling and the push-over study. In this model, springs in top and bottom planes are included at the center of gravity of each section, by distributing couplings modelling a master central node connected with the spring and several slave nodes in the shell section. External loads are applied in a similar way using kinematics linear couplings.

3 Seismic loads

The design seismic motion is defined by the response spectra of the horizontal (figure 8) and vertical components, considering a modal damping ratio of 5%, and with four levels of the peak acceleration shown in table 1, with the associated return period and the probability of not exceeding them during the life of the structure of 200 years.

For the nonlinear dynamic analysis, the Northridge 1994 N-S (figure 9) and E-W earthquake components were initially used. A comparison of the pseudo-acceleration response spectra of this earthquake and the correspondent design spectra for the ULS state is presented in the same figure.

Using the program Simqke [5] a set of artificial quakes adjusted with each of the pseudo-acceleration design spectrums with low tolerances were developed. Figure 10 shows one artificial earthquake named “71” that adjust on the security side of the design PSA spectra for the ULS state.

These earthquakes were applied first as uniform base excitation imposing their accelerations at all nodes connected with terrain. After that, the possibility of non-uniform support movement [6] was analyzed imposing different quake displacements in each support, using the same earthquake but with temporal translations as a function of the Vs waves velocities.

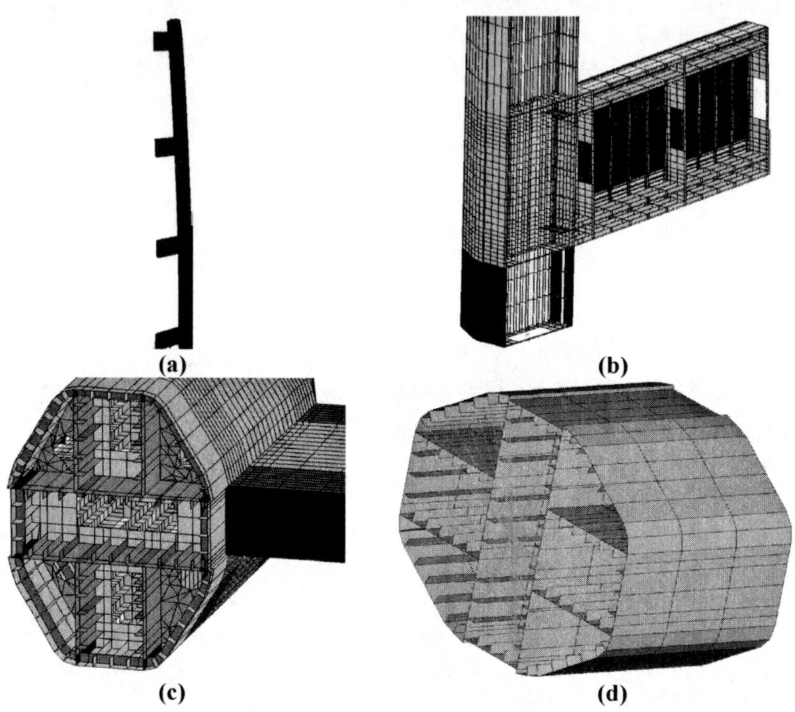

Figure 7: (a) Messina tower local shell model; (b) Section detail of one connection; (c) Column stiffeners; (d) Kinematics couplings for load application.

Table 1: Peak ground acceleration levels in each load condition.

	SLS1	SLS2	ULS	SILS
Peak ground acceleration (m/s^2)	1.2	2.6	5.7	6.3
Return Period (years)	50	200	2000	10000
% Probability of not exceeding	1.8	37	90	98

4 Vibration modes

Considering the global bridge model, table 2 shows the first six modes calculated by different analysis. In the case of column named "nonlinear", modes (Φ) and frequencies (ω) are calculated using the full nonlinear stiffness matrix (K_{NL}) obtained from an initial nonlinear static step with dead loads:

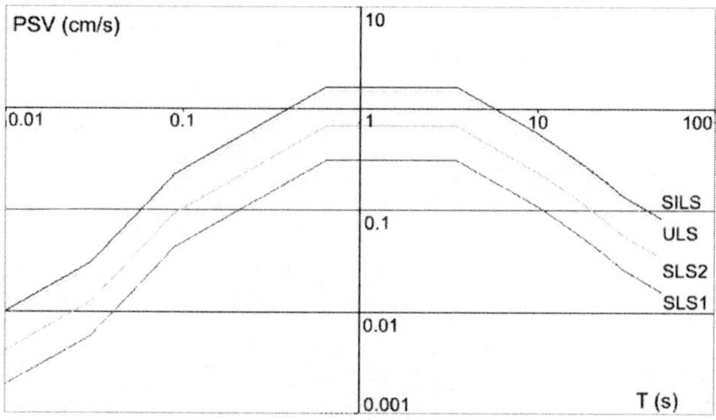

Figure 8: Pseudo-velocity spectrums.

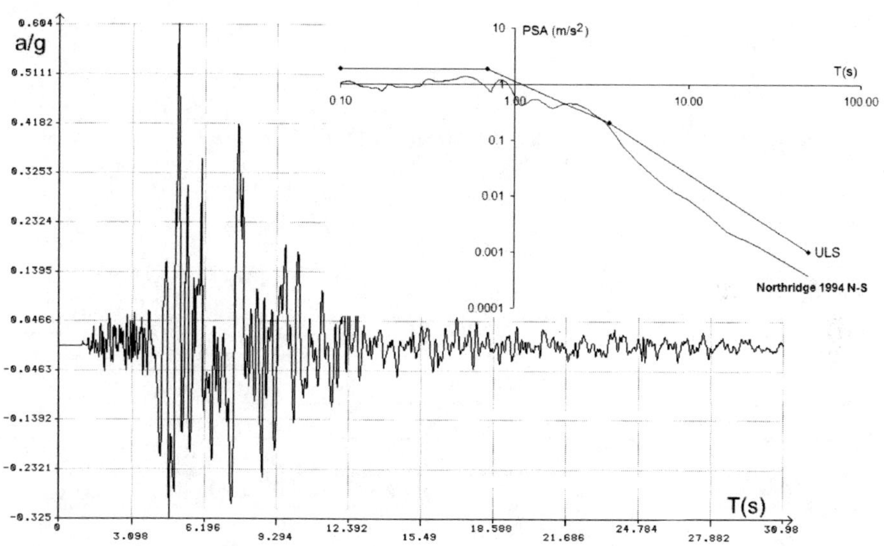

Figure 9: Northridge (1994) N-S time history quake and PSA spectra.

$$\begin{aligned} (K_{NL} - \omega^2 M)\Phi = 0 \\ K_{NL} = K_L + K_\sigma + K_G \end{aligned} \tag{1}$$

where $\boldsymbol{K}_{\mathbf{L}}$ is the linear stiffness matrix, $\boldsymbol{K}_{\boldsymbol{\sigma}}$ is the geometric stiffness due to initial stresses, $\boldsymbol{K}_{\mathbf{G}}$ is the nonlinear stiffness due to node displacements, and $\mathbf{M}$ is the lumped mass matrix.

Values in the column named "live loads" are calculated with the full $\boldsymbol{K}_{\mathbf{NL}}$, obtained form a second step that adds to the initial dead loads the loads due to four centered trains with 750 m of length. The variations of vibration modes are significant but less than 10% of the previous values. Finally, in the column

"Only K_σ", vibration periods calculated using only the geometric stiffness matrix from dead loads are shown. The variations in modes are less than 5% of the initial values, although the modes 5 and 6 change their positions.

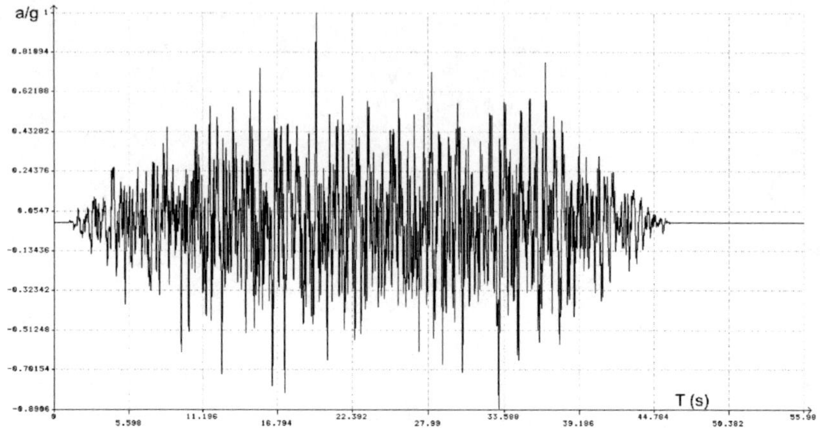

Figure 10: Artificial earthquake 71 time history and PSA spectra.

Table 2: First six vibration periods (s) of bridge global model with different analysis.

Mode	Nonlinear	Only K_σ	Live loads	Description
1	31.98	32.18	32.81	First horizontal symmetrical
2	17.48	17.57	19.21	First horizontal anti-symmetrical
3	16.54	16.55	17.31	First vertical anti-symmetrical
4	12.28	12.10	12.62	First vertical symmetrical
5	12.01	12.31	12.32	First torsional anti-symmetrical
6	11.95	12.02	11.51	Second horizontal symmetrical

In the case of local Messina tower model of figure 6a, the first transversal (T_1 = 3.26 s) and longitudinal (T_2 = 2.97 s) modes, obtained in the nonlinear method are shown in figure 11, next to the first mode of the global model. To achieve these periods about 3 s. in the global model, more than 50 modes must be calculated due to the great number of vibration modes involving main cable vibrations that appear. For that reason, to reach mass participating factors over the 90% in each spatial direction, a great number of modes (more than 100) must be considered in the global model.

5 Dynamic nonlinear analyses

Due to the geometrically nonlinear behavior, all the global seismic analyses were done by direct integration of coupled dynamic equilibrium equations using

Newark's method with average acceleration integration ($\gamma = 0.5$, $\beta = 0.25$), to achieve stability unconditionally.

Rayleigh damping was considered by adjusting modal damping of several pairs of modes to check its influence in the results. Figure 12 shows the pairs of modes used. Finally the modes 1 and 166 were used ($T_1 = 31.98$ s, $T_{166} = 1.11$ s) with a damping coefficient for stiffness matrix of 0.045, and 0.017 for mass matrix.

Seismic analysis was carried out in two steps; the first to apply dead and live loads and then the earthquake load was applied with the response integrated with time steps of 0.02 s. Considering the artificial earthquake of figure 10, applied in longitudinal and transversal direction, figure 13 shows several nodal time history displacements, and figure 14 shows the variation of longitudinal moment and axial force in one column of the Messina tower for the ultimate limit state ULS71, calculated with the envelope of seismic results.

$$ULS71 = 1.15PP + 1.5PN + 1IMP + 1.1QA + 1VS + 1VT$$

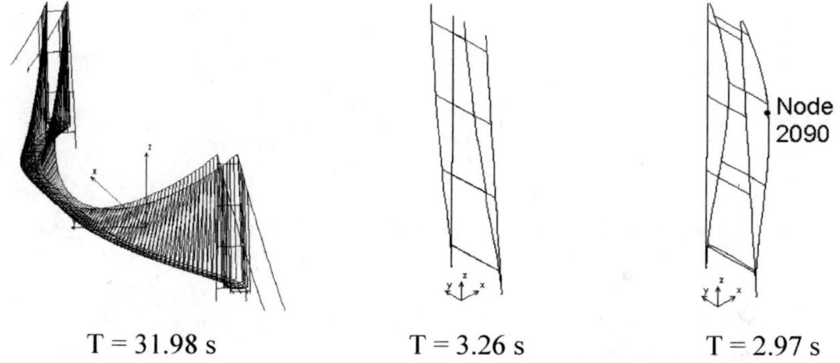

Figure 11: Vibration modes in global and local models.

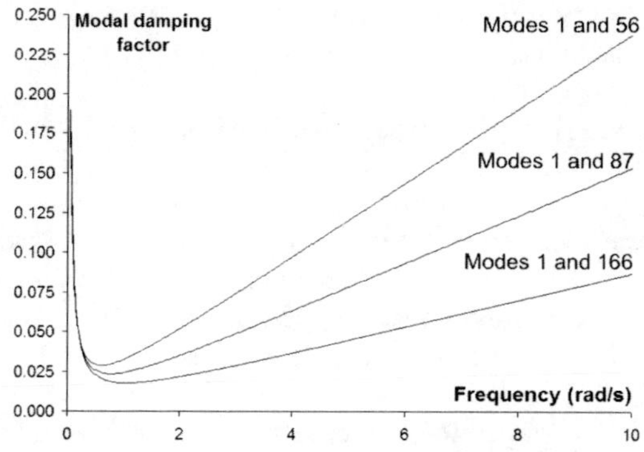

Figure 12: Rayleigh damping curves.

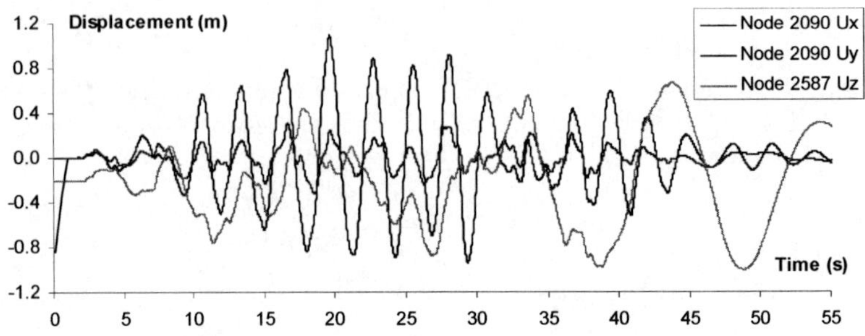

Figure 13: Time history of nodal displacements.

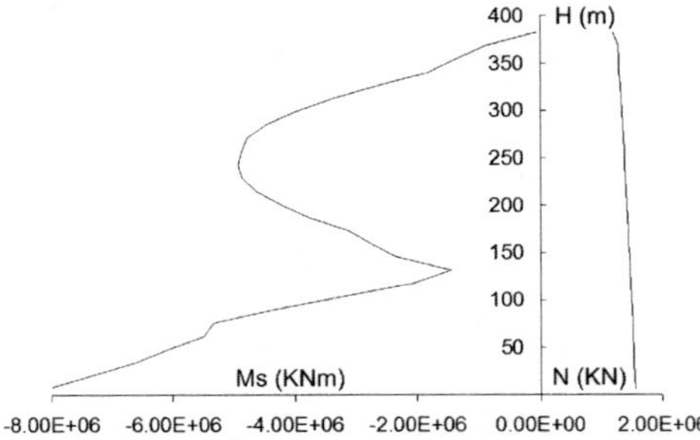

Figure 14: Longitudinal moment and axial load in Messina tower shaft for ULS 71.

Tower maximum displacements in longitudinal direction are about 1.1 m (see position of node 2090 in figure 11). Transversal earthquake displacements in towers are smaller, with maximum values about 0.5 m, located in the upper part. In the central position of the bridge (node 2587 in figure 13), maximum seismic displacements are smaller in comparison with the rest of loads, with maximum values of 0.42 m. in longitudinal direction, 1.29 m. in transversal direction, and 1 m in vertical displacements.

6 Messina tower longitudinal push-over analysis

A nonlinear static analysis including geometrical and material nonlinearities was carried out using the local shell model of Messina tower shown in figure 6. The aim of the push-over analysis is to determine, statically, the resistant capacity, ductility, collapse mechanism and security level of the structure, without performing complex nonlinear seismic analysis [7, 8]. The ultimate state considered was: ULS71 = 1.15PP + 1.5PN + 1IMP + 1.1QA + LF x VS+ 1VT.

In the first nonlinear stage all loads except the seismic ones were applied, and in the second step seismic load was applied proportionally until collapse. Several methods have been proposed to generate the equivalent seismic load pattern in the push-over analysis [9]. In our case, since the longitudinal seismic response of the tower is determined by the second local mode of vibration ϕ_2 (T = 2.97 s in figure 11), the load pattern (P_s) used (figure 15a) was calculated using the tower masses associated to dead loads (m) in each section with:

$$P_s = \lambda mg\, \phi_2 \tag{2}$$

where the load factor λ was calculated in a way that the static application of load Ps gives a base moment equal to the maximum obtained with the correspondent seismic time history analysis.

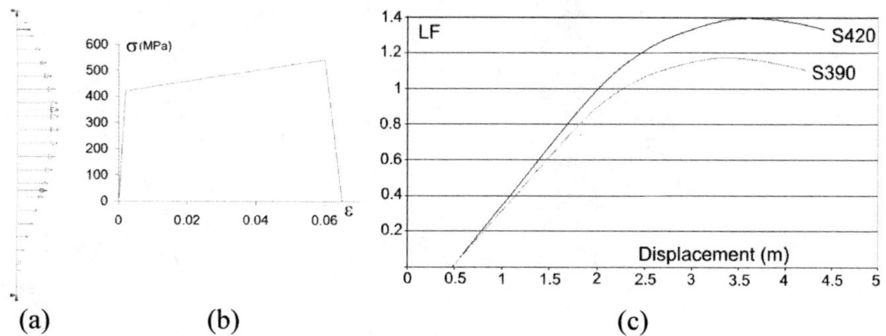

Figure 15: (a) Seismic load pattern; (b) Stress-strain curve for steel S420; (c) Longitudinal pushover results.

Steel types S390 and S420 were considered in the analysis, using stress-strain diagrams like the one shown in figure 15b for the S420, with an elastic modulus E of 2.1 x 10^8 KPa, initial plastic deformation of 0.2%, elasticity tangent modulus of E/100 and ultimate deformation of 6%.

In stage 1, without seismic loads, maximum displacements are 1.18 m in tower head, and 0.48 m in node 2090 (see position in figure 11); the entire tower was in elastic state. Figure 15c shows the longitudinal displacements obtained in node 2090 during the second nonlinear stage with linear increments in the load factor LF that multiplies the equivalent static load Ps. With S390 steel, plastic behavior starts for LF = 0.94, and the collapse load factor takes place for LF = 1.17 and a longitudinal displacement of 3.4 m in node 2090. Using the S420 steel, global ductility grows and the maximum load factor is about 1.4.

7 Conclusions

- Several seismic preliminary checks of the initial design project of the Messina Bridge have been presented.
- The use of simplified global models with beam and springs elements allows to perform geometrically nonlinear static and seismic time history analysis.
- Results obtained from global models are transferred to detailed shell local models, which allowed us to rerun analysis including geometrical nonlinearities and non-linear materials. These local models, combined with normative checks, help us to understand the structural behaviour under collapse and the position and problems of critical areas.

— Several analysis must be improved, such as including material nonlinearities in the global beam bridge model, seismic effects of tuned mass dampers located in towers, and consideration of residual stresses and local imperfections in local shell models.

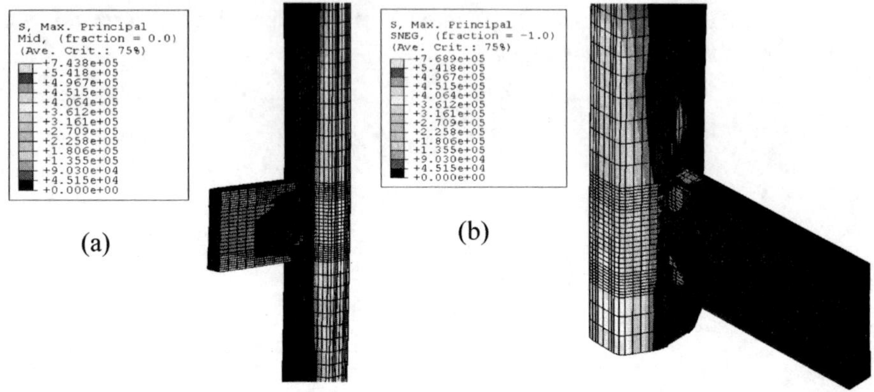

Figure 16: Maximum principal stresses previous to the collapse in the proximities of node 2090 (a), and in the base tower (b).

References

[1] Cosmos/m v.2.9. Finite Element Analysis System: I) User Guide, IV) Advanced Modules. Structural Research & Analysis Corp., (2004).

[2] Abaqus, I., (ed.) Abaqus 6.5. Standard Uses Manual. Pawtucket, Rhode Island, 2004.

[3] Bathe, K.J., Finite Element Procedures. Prentice Hall: Englewood Cliffs, N.J., 1996.

[4] Crisfield, M.A., Non-Linear Finite Element Analysis of Solids and Structures. Vol. 2: Advanced Topics. John Wiley & Sons: Chichester, 2001.

[5] SIMQKE, A Program for Artificial Motion Generation. NISEE/Computer Applications (1986).

[6] Chopra A.K., Dynamic of Structures. Theory and Applications to Earthquake Engineering. Pearson Prentice Hall: New Jersey (2007).

[7] Lu, Z., Ge, H. & Usami, T., Applicability of Push-over Analysis Based Seismic Performance Evaluation Procedure for Steel Arch Bridges. Engineering Structures 26 (2004).

[8] Archer, G.C., A Constant Displacement Iteration Algorithm for Nonlinear Static Push-Over Analyses. Electronic Journal of Structural Engineering, 2 (2001).

[9] Endo, T.K., Chihiro, K., Shigeki, U., Analytical Study on Seismic Performance Evaluation of Long-Span Suspension Bridge Steel Tower. 13th World Conference on Earthquake Engineering, Vancouver. Canada (2004).

Recent tendencies in wind storm climatology with implications to storm surge statistics in Estonia

Ü. Suursaar[1], J. Jaagus[2] & T. Kullas[1]
[1]*Estonian Marine Institute, University of Tartu, Estonia*
[2]*Institute of Geography, University of Tartu, Estonia*

Abstract

The aim of this paper is to present an overview of Estonian storm climatology and a statistical study of wind storms and storm surges in the coastal waters of Estonia, Baltic Sea. The results show a general positive trend both in local storminess and storm surge heights over the last century. These tendencies are probably associated with increased westerly circulation and cyclonic activity in the Northern Atlantic. While in Estonian tide gauges, the mean sea level rise (1.5–2.7 mm/yr) is roughly equal to or insignificantly higher than the recent global sea level rise estimates (1.5–1.7 mm/yr), the trend estimates for annual maximum sea levels vary between 3 and 10 mm/yr for different tide gauges and study periods. Maximum value analysis revealed that, in general, the empirical data of both annual maximum wind speeds and sea levels follow the Gumbel distribution. However, in some windward bays of western Estonia two storm surge events (in 1967 and 2005) are inconsistent with the theoretical distributions, which indicate that, in these locations, the most extreme sea level events are hardly predictable by means of return statistics.
Keywords: climate change, storm surges, hurricanes, sea level, trends.

1 Introduction

Estonia lies in the zone of moderate latitudes (57.5°–59.5° N, Fig. 1), which serves as a prolongation of the so-called North Atlantic storm track (e.g. Soomere [1]). As such, the predominant meteorological conditions in the area are characterized by the frequent passage of cyclones, which accordingly cause considerable sea level fluctuations in the nearly tideless eastern section of the

WIT Transactions on Engineering Sciences, Vol 58, © 2007 WIT Press
www.witpress.com, ISSN 1743-3533 (on-line)
doi:10.2495/EN070051

Baltic Sea. During the last half-century, significant changes in the regional wind regime are evident. Many authors [2,3] have reported increases in both the winter westerly wind speed component and in the winter North Atlantic Oscillation (NAO) index. Also, cyclonic activity has risen both above the North and Baltic Seas and vitalization of coastal processes have been observed in Estonia by Orviku *et al* [4]. During severe storms in Estonia, the highest economic losses are concentrated in the coastal zone and the worst damages occur mainly due to storm surges. On 7–9 January 2005, the new highest recorded storm surge (275 cm) occurred in Pärnu, Estonia, and rise in storm-related risks in the future is foreseen [5,6].

This paper looks at the long-term changes both in wind storm and extreme sea level events. Our aim is also to investigate how changes in wind speed and direction affect sea level in the study area and to discuss the possibility of estimating maximum expected future surge heights in Estonian coastal waters.

2 Material and methods

2.1 Wind and storminess

At the present time, the network maintained by the Estonian Meteorological and Hydrological Institute (EMHI) includes 21 meteorological stations.

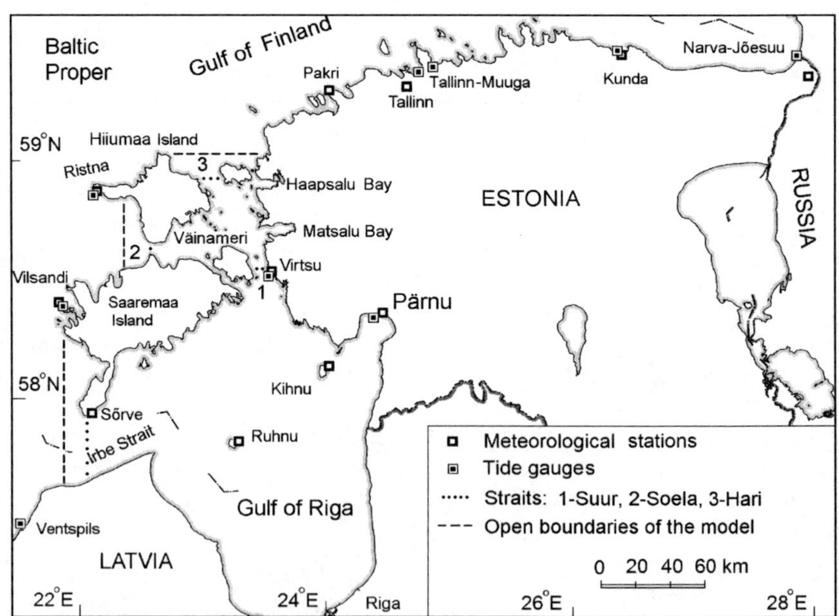

Figure 1: Study area with selected meteorological and tide gauge stations.

Wind data from Kihnu and Vilsandi stations (Fig. 1) were used in this study. The latter has the most open location for wind measurements among all Estonian

stations and features by the highest long-term mean wind speed (6.6 m/s). The stations are equipped with automatic weather stations, which provide data sets of hourly average and gust wind speeds, as well as wind directions.

Historically, time series for wind speeds start in 1855 (at Pakri). Observations started in 1865 at Vilsandi, and in 1931 at Kihnu. However, some changes in methodology and equipment, as well as gaps in data series, considerably abridge the extension of applicable data. Monthly and annual numbers of storm days at Vilsandi station (over the period 1948–2005) and Kihnu (1950–2005) were sorted out for this study. Decadal variations in storm frequency and annual maximum wind speed were studied. A storm was defined when the mean wind speed during a single observation (10 minutes) was 15 m/s or higher. A storm day was defined when storms were recorded during at least one observation. The frequency of storms (i.e., the number of storm days per year) was used to measure storminess. The datasets additionally include information about storm intensity, such as the highest mean wind speed, wind direction and duration.

At the stage of pre-treatment, an attempt was made to ensure homogeneity of the data sets of storms. In general, local long-term wind time series can be sensitive for building activity and changes in vegetation near the stations, but also to inhomogeneity caused by instrument changes. An important change from weathercocks to automatic anemorhumbometers took place at Vilsandi and Kihnu on November 1976. The corresponding corrections could be found from professional handbooks. For example, 15 m/s mean wind speed corresponds to previous 17 m/s and 26 m/s equals to previous 30 m/s. Secondly, the interval of measurements has decreased from 6 hours (prior to 1965) to 3 h, and for the few last years, hourly data is available. In this regard, data about storm duration was used. For example, if there are two consecutive storm wind readings, a storm day can be equally defined both for 6 h and 3 h intervals. But the storm day, which is defined by one measurement in case of 3 h interval, yields just 0.5 storm days on the basis of 6 h interval.

Extreme value analysis was performed by calculating empirical return periods and fitting theoretical distributions to the samples of annual maximum values.

2.2 Sea level

The EMHI operates 12 tide gauges. Automatic tide gauges, which provide hourly data, are currently located at Pärnu, Narva-Jõesuu and Ristna (Fig. 1). We used the databases of monthly mean and extreme sea levels at Pärnu and Tallinn. They represent relative sea level values in regard to the Kronstadt zero benchmark, which is nearly equal to the long-term mean sea level for the Estonian coast. Pärnu has the highest sea level variability among the tide gauges, according to Suursaar *et al* [5]. In Tallinn, the capital city of Estonia, regular sea level measurements started in 1809. The near continuous data sets are available from 1842, but the measurements were discontinued in 1996 due to construction work at the Port of Tallinn. After that, a mareograph was installed to the nearby Port of Muuga.

Tendencies in sea level time series were analysed using linear regression analysis. Taking into account the gaps, the series for extreme value analysis included 83 values at Pärnu (in 1923–2005) and 85 values in Tallinn (in 1899–1995). In addition to statistical analysis, some sensitivity analyses were carried out using a two-dimensional (2D) hydrodynamic model, which simulates sea level variations depending on wind stress and boundary sea level forcing. The model domain includes both the Gulf of Riga and the Väinameri sub-basins (Fig. 1) with a 1 km grid resolution, yielding in total 18,964 marine grid-points. For a more detailed model description, see the hindcast studies by Suursaar *et al* [5,7].

3 Results and discussion

3.1 Tendencies in storminess and their relations to large-scale circulation

The results show a clear increase by trend both in frequency of storms (Fig. 2ac) and in annual maximum wind speeds (Fig. 2cd). As an average, the number of storm days increased from 16 to 18 at Vilsandi and from 8 to 15 at Kihnu. The increase in maximum wind speed amounted from 19.5 to 22 m/s at Vilsandi and from 18 to 22.5 at Kihnu. The trends for Kihnu are statistically significant on $P<0.05$ level. The presented trends in storm frequency are not as steep as shown in our preliminary analysis for the period of 1950–2002 [4] for two reasons. Firstly, the previous study did not fully take into account the different measuring intervals and the data were a bit understated prior to 1965. Secondly, storminess experiences certain natural cycles with a period of about 35–40 years and the last added years belong to a decreasing phase of this periodicity (Fig. 2). Similar periodicity appears also in the winter NAO index (Fig. 2e), in sea level records both in Estonia (Fig. 3) and in neighbouring countries (e.g. Wakelin *et al* [8]). According to studies by Jaagus and co-authors (e.g. [4,5]), the rise in storminess appears mainly in winter months, which is connected with the stronger zonal circulation (i.e. westerlies, the NAO) in the same months.

3.2 Tendencies in sea level

Time series of annual mean sea levels show slightly increasing tendencies (Fig. 3) with rise rates between 0.1 and 1.0 mm/yr in Estonia. With regard to the relative sea level of a location, the two main factors influencing its variations are global sea level change and vertical land movement, which is +1.5 mm/yr at Pärnu and +1.8 mm/yr at Tallinn according to Vallner *et al* [9]. After adjusting the sea level rates to account for land uplift rates, we calculated sea level rise rates of 1.9 mm/yr in Tallinn (1842–1995) and 2.6 mm/yr in Pärnu (1924–2005). The estimates for the last 50 years are 1.6 mm/yr at Tallinn and 2.7 mm/yr at Pärnu (Fig. 3ab). However, the trend estimates for annual maximum sea levels vary between 3 and 10 mm/yr for different tide gauges (Fig. 3cd).

Except in Pärnu, the mean sea level rise rates in Estonian tide gauges are roughly equal to or insignificantly higher than the most recent global sea level rise estimates, which are around 1.7 mm/yr according to Church and White [10].

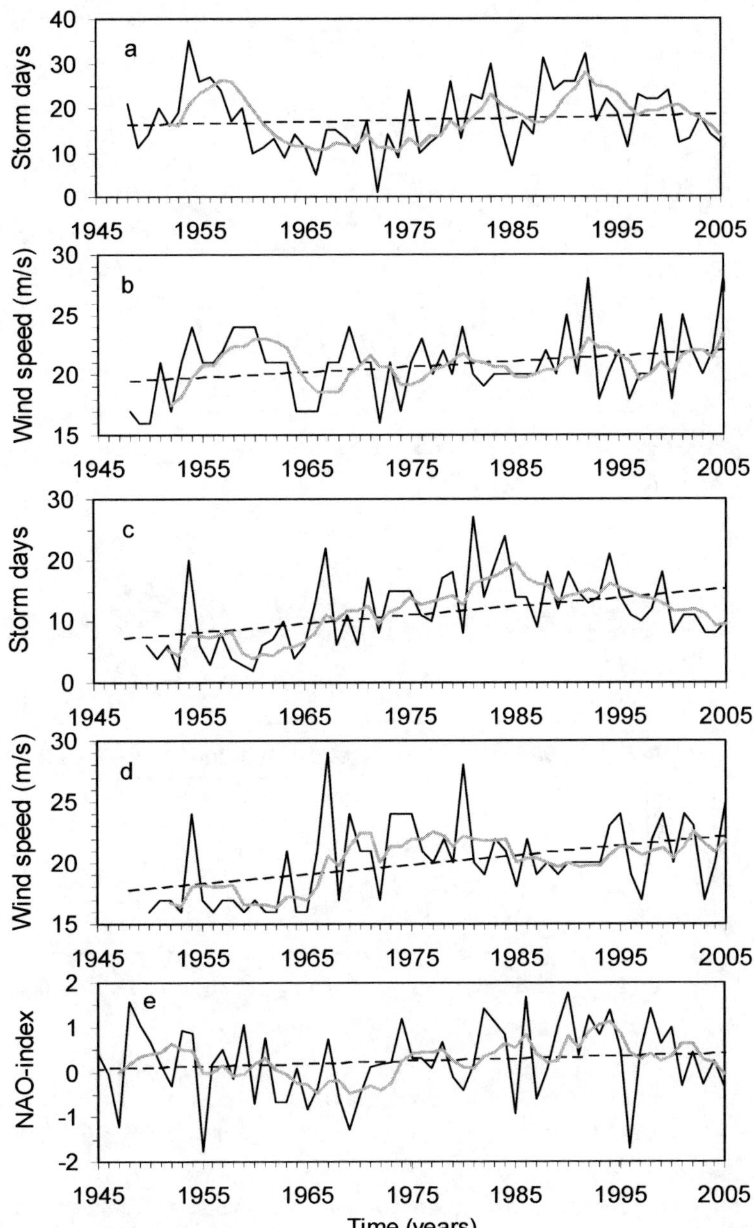

Figure 2: Decadal variations in annual number of storm days at Vilsandi (a) and Kihnu (c), annual maximum wind speeds at Vilsandi (b) and Kihnu (d); variations in the November to March NAO index (e). 9-year moving averages and linear trendlines are added to the original series.

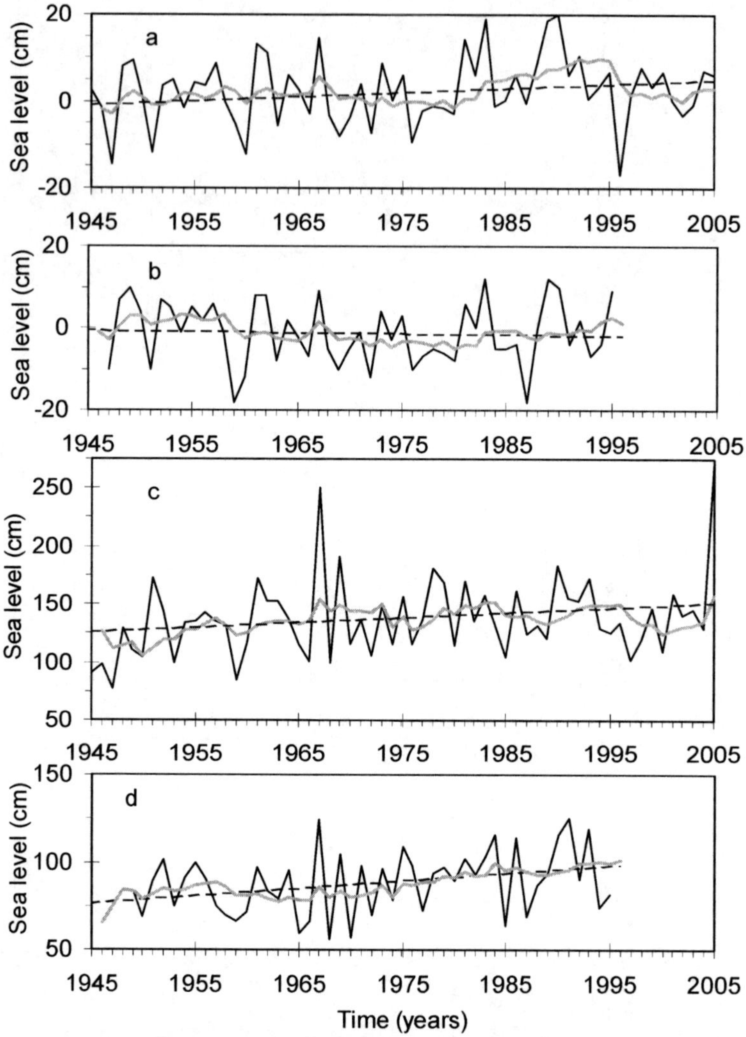

Figure 3: Decadal variations in annual mean relative sea level at Pärnu (a) and Tallinn (c), annual maximum sea levels at Pärnu (b) and Tallinn (d) together with 9-year moving averages and linear trendlines.

Pärnu's positive sea level trends in annual time series appear due to the significantly more positive trends in winter sea level, since during the summer, such trends are less steep or even negative [5]. The significantly higher mean sea level rise in winter correlates with increased local storminess during the same months and with the greater intensity of westerlies in winter, as described also by the NAO index (Figs. 3,2e). The excessive Pärnu sea level rise rate over the global estimates can be explained by hydrodynamic mechanisms, and will be discussed further below.

3.3 Tendencies in maximum values and the sensitive coastal areas

While maximum recorded gust wind speed is 48 m/s in Estonia (at Ruhnu in 1969), the maximum corrected sustained wind speeds reach about 30 m/s at several stations (Figs. 2,4cd). To find the empirical return periods (Fig. 4), the events were ranked and the observation period was divided by the number of events. Using the least-squares method, mode (a) and scale parameters (b) for theoretical Gumbel distributions were calculated for the stations in question. The corresponding values were 20.1 and 2.1 for Vilsandi, 19.2 and 2.5 for Kihnu, 120.5 and 27.1 for Pärnu, and 76.3 and 14.1 for Tallinn.

The plots against the empirical data show a more or less satisfactory fit in case of maximum wind speeds (Fig. 4cd) and for the sea level (e.g. Fig. 4b) other than sea level at Pärnu (Fig. 4a). Probably no extreme value distribution could predict the two extreme sea level events of 253 cm in 1967 and of 275 cm in 2005. Even before 2005, 253 cm was considered as an outlying value with a theoretical recurrence period of some 300–1000 years, but it was repeated and surpassed just 38 years later. The graph (Fig. 4a) is similar to the graphs for gust wind speed that are contaminated by occasional tornados [11], or sea level data that includes tsunami events.

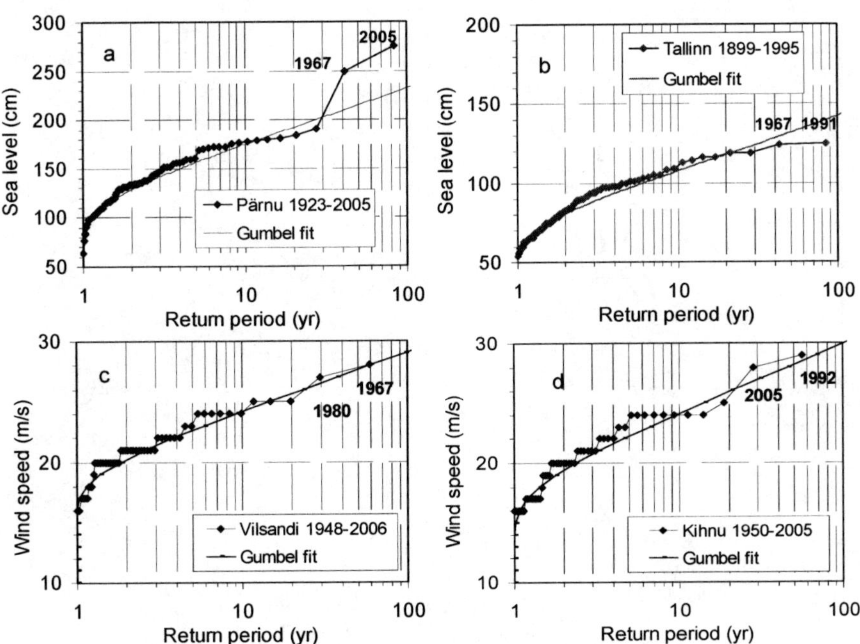

Figure 4: Return periods based on annual maximum sea level data at Pärnu (a) and Tallinn (b), and maximum sustained wind speeds at Vilsandi (c) and Kihnu (d), together with corresponding Gumbel distributions. The years with the highest values are marked for each graph.

The two anomalous values at Pärnu were nevertheless caused by normal storms. It appeared from a modelling study [5] that due to the specific configuration of the Gulf of Riga and Pärnu Bay, the sea level is proportional to the wind speed in the power of 2.4 at Pärnu Bay. It means that at the upper range of wind speeds, a slight incremental increase yields a progressively higher incremental increase in storm surge level (Fig. 4a). Using a 2D hydrodynamic model, we searched for such highly sensitive locations to wind storm conditions.

To begin, we identified the maximum sea level heights for each grid point of the model domain at a constant wind speed (Fig. 5c). Next, the corresponding wind directions yielding these maximum values were found. The patterns of wind directions resemble tidal charts, particularly in the northern sub-basin of the Väinameri (Fig. 5a), where the phase (wind direction) rotates clockwise around an amphidromic point with zero-amplitude. Fig. 5 demonstrates that the most sensitive areas for wind forcing are the shallow windward bays of Haapsalu, Matsalu and Pärnu. The most effective wind direction for these bays is 220°–240° (SW).

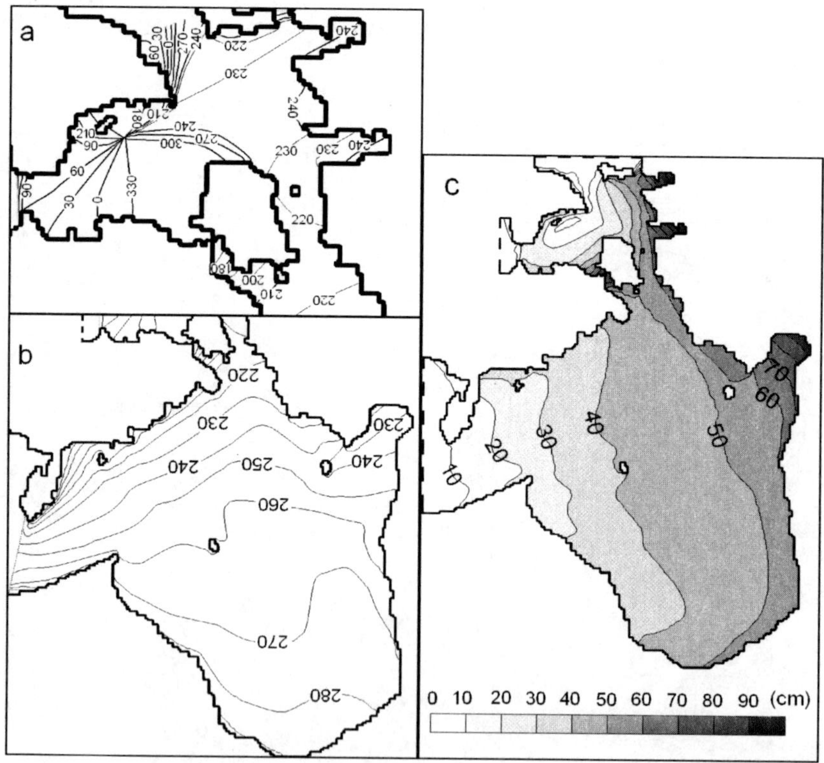

Figure 5: Horizontal distribution of the highest possible sea levels (above the Baltic background sea level) modelled with 20 m/s stationary wind (c); the corresponding wind directions that produce the highest sea level of a location at Väinameri (a) and the Gulf of Riga (b).

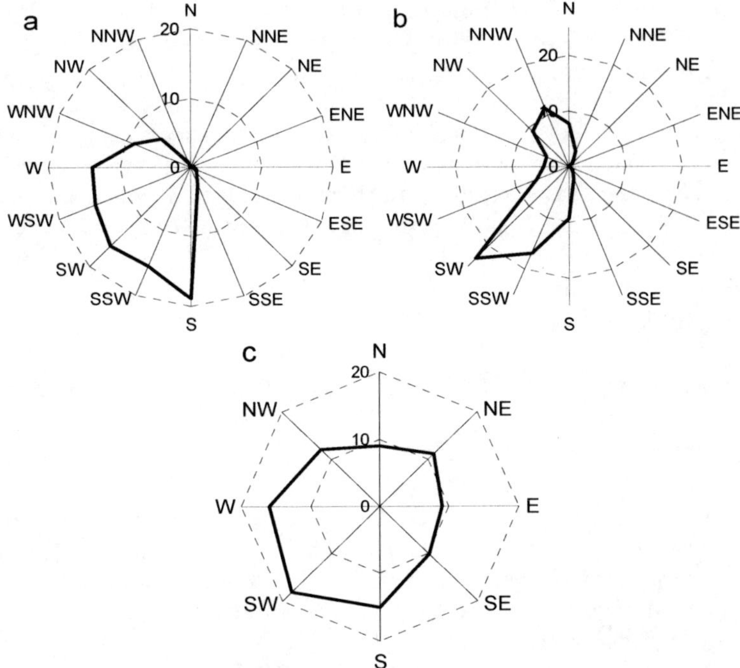

Figure 6: Directional distribution of storm wind events (winds >15 m/s) at Kihnu (a) and Vilsandi (b) compared to traditional wind rose for all wind directions at Kihnu (c) over the period 1950–2005.

SW is also the most frequent storm wind direction (Fig. 6), as well as the direction where the strongest winds can blow [1,5]. Because of the specific configuration and the exposure of bays to the direction of the strongest possible winds (Figs. 5,6), the sea level regime is very sensitive to changes in wind condition. As a result, in case of decadal trend in wind conditions (Fig. 2; [2,3]) the sea level change rates of a semi-enclosed basin can deviate from the global estimates (Fig. 3). A positive trend in wind speed and storminess should result in a steeper than average sea level trend on the windward side and one that is less steep on the leeward side.

4 Conclusions

It appeared from the statistical analysis that a positive trend both in local storminess and storm surge heights over the last century is visible in Estonia. These tendencies are associated with increased westerly circulation and cyclonic activity in the Northern Atlantic. At Estonian tide gauges, the mean sea level rise rates (adjusted to take into account the land uplift rates) are 1.5–2.7 mm/yr over the last century and the trend estimates for annual maxima vary between 3 and 10 mm/yr. The trend in the windward-located shallow and narrow Pärnu bay is steeper than predicted by the mean global sea level rise estimate and the storm surges in 1967 and 2005 are inconsistent with the theoretical distribution. The excessive rise in local sea level can be explained by the local sea level response to the changing regional wind climate and intensification of cyclones.

The statistical study of storm winds and storm surges can be useful for shipping companies and port authorities, as well as for spatial planners and stakeholders in the coastal regions, which are threatened by wind storms and sea level rise. Using hydrodynamic modelling, such sensitive regions were defined and storm surge mechanisms studied.

Acknowledgements

The study was supported by the ESF grant projects No. 5763, 5786 and 5929.

References

[1] Soomere, T., Extreme wind speeds and spatially uniform wind events in the Baltic Proper. *Proc. Estonian Acad. Sci. Eng.*, **7**, pp. 195–211, 2001.

[2] Siegismund, F. & Schrum, C., Decadal changes in the wind forcing over the North Sea. *Clim. Res.*, **18**, pp. 39–45, 2001.

[3] Alexandersson, H., Schmidt, T., Iden, K. & Tuomenvirta, H., Long-term variations of the storm climate over NW Europe. *The Global Atmos. Ocean. Syst.*, **6**, pp. 97–120, 1998.

[4] Orviku, K., Jaagus, J., Kont, A., Ratas, U. & Rivis, R., Increasing activity of coastal processes associated with climate change in Estonia. *Journal of Coastal Research*, **19**, pp. 364–375, 2003.

[5] Suursaar, Ü., Jaagus, J. & Kullas, T., Past and future changes in sea level near the Estonian coast in relation to changes in wind climate. *Boreal Environment Research*, **11**, pp. 123–142, 2006.

[6] Lowe, J.A., Gregory, J.M. & Flather, R.A., Changes in the occurrence of storm surges around the United Kingdom under a future climate scenario using dynamic storm surge model driven by the Hadley Centre climate models. *Climate Dynamics,* **18**, pp. 179–188, 2001.

[7] Suursaar, Ü. & Kullas, T., Influence of wind climate changes on the mean sea level and current regime in the coastal waters of west Estonia, Baltic Sea. *Oceanologia*, **48**, pp. 361–383, 2006.

[8] Wakelin, S.L., Woodworth, P.L., Flather, R.A. & Williams, J.A., Sea-level dependence on the NAO over the NW European Continental Shelf. *Geophys. Res. Lett.*, **30**, Art. No. 1403, 2003.

[9] Vallner, L., Sildvee, H. & Torim, A., Recent crustal movements in Estonia. *J. Geodyn.*, **9**, pp. 215–223, 1988.

[10] Church, J.A. & White, N.J., A 20th century acceleration in global sea-level rise. *Geophys. Res. Lett.*, **33**, L01602, 2006.

[11] Cheng, E. & Yeung, C., Generalized extreme gust wind speeds distributions. *J. Wind Engin. Ind. Aerodyn.*, **90**, pp. 1657–1669, 2002.

Safety design of lightweight roofs exposed to snow load

M. Holický
CTU in Prague, Klokner Institute, Prague, Czech Republic

Abstract

The collapse of a number of lightweight roofs during the winter period of 2005/2006 initiated an international discussion concerning the reliability of roofs exposed to permanent snow load. In some countries, all available measurements of snow loads have been newly evaluated and relevant standards have been promptly revised. Newly developed maps of snow loads are based on the principles of the European standards specifying the characteristic value of snow load on the ground as the 98 fractile of annual extremes (50 year return period). This paper provides a critical analysis of presently accepted design procedures taking into account available data of snow load. In the reliability analysis permanent load is described by normal distribution, snow load by Gumbel distribution and resistance by lognormal distribution. It appears that the partial factor design method provided in the present European standards may not guarantee an adequate reliability level of lightweight roofs. An alternative procedure for the safety design of roofs exposed to self weight and snow load only is therefore proposed.
Keywords: lightweight roofs, snow load, safety, design.

1 Introduction

The reliability of lightweight roofs has become an important topic of structural design, particularly after the winter period 2005/2006 when a number of roofs in Europe collapsed. In some countries, available measurements of snow loads have been newly evaluated and relevant standards have been promptly revised. Newly developed maps of snow loads take into account the principles of valid European standards [1–3] specifying the characteristic value of snow load on the ground as the 98 fractile of annual extremes. The design value of snow load is then determined using the partial factor 1,5.

WIT Transactions on Engineering Sciences, Vol 58, © 2007 WIT Press
www.witpress.com, ISSN 1743-3533 (on-line)
doi:10.2495/EN070061

This paper provides a critical analysis of presently accepted design procedures taking into account available measurements. Normal distribution for permanent load, Gumbel distribution for snow load and lognormal distribution for resistance are assumed in the reliability analysis [4,5]. The presented study is an extension of the recent paper [6] and confirms its conclusion that available European standards may not guarantee an adequate reliability level for lightweight roofs. An alternative procedure of the safety design of lightweight roofs exposed to permanent and snow load only is, therefore, proposed.

2 Partial factor design

In accordance with the principles of the present suite of European standards (Eurocodes) the characteristic value of the snow load on the ground s_k is specified as the 0,98 fractile of annual extremes (50 years′ return period) [1,2] assuming the Gumbel distribution. The characteristic load on the roof is then determined as

$$s_{s,k} = \mu C_{\mathrm{e}} C_{\mathrm{t}} s_k \tag{1}$$

where μ denotes the shape factor (for horizontal roofs equal to 0,8). C_{e} and C_{t} denote the exposure and thermal factors considered usually as unity [3] (and omitted further on). In the design of a structural component exposed to the permanent load G and snow load S, the value of a generic resistance R is determined using the partial factor method and the fundamental load combination [1] as

$$R_{\mathrm{k}} / \gamma_{\mathrm{M}} = \gamma_G G_{\mathrm{k}} + \gamma_Q s_{\mathrm{s,k}} \tag{2}$$

Here R_{k} denotes the characteristic resistance (0,05 fractile), γ_{M} the resistance partial factor (considered by a generic value 1,15), γ_G the partial factor of permanent load (considered by a recommended value 1,35), G_{k} the characteristic value of the permanent load (the mean of G), γ_Q the partial factor of the snow load (considered by a recommended value 1,5). Note that equation (2) corresponds to the fundamental load combination given in EN 1990 [1] by expression (6.10). The other (more economic) load combinations provided in [1] are not considered here but may be treated in a similar way.

3 Basic variables of reliability analysis

The reliability analysis of a structural member exposed to a permanent load G and snow load S is, based on the limit state function g($\boldsymbol{X}$), given as

$$\mathrm{g}(\boldsymbol{X}) = K_{\mathrm{R}} R - K_{\mathrm{E}} (G + S_{50}) \tag{3}$$

The basic variables $\boldsymbol{X}$, entering equation (3), are described in Table 1. In accordance with EN 1990, the design life time of 50 years is considered and, therefore, the 50 years′ extreme S_{50} of the snow load $s_{\mathrm{s,k}}$ on the roof (considering the shape factor μ = 0,8) is assumed in equation (3). Note that the statistical characteristics given in Table 1 are derived taking into account the principles of Eurocodes [1–3].

In the following analysis, the load ratio χ is given as a fraction of the characteristic value of the snow load $s_{s,k}$ and the total load $G_k + s_{s,k}$

$$\chi = s_{s,k} / (G_k + s_{s,k}) \tag{4}$$

The load ratio χ for lightweight roofs is expected within the interval from 0,4 to 0,8. For a given χ and $s_{s,k}$ the characteristic permanent load follows from equation (4) as

$$G_k = s_{s,k} (1 - \chi) / \chi \tag{5}$$

The characteristic value $s_{s,k}$ is considered by the normalized value 1 (say 1 kN/m^2). The mean of S_{50} is approximately equal to the characteristic value $s_{s,k}$ (see Table 1).

Alternatively, the partial factor of snow load γ_Q is proposed as a quantity dependent on the load ratio χ (similarly as suggested in recent studies [7,8] for the partial factor of variable actions)

$$\gamma_Q = 1 + \chi \tag{6}$$

It appears that the partial factor γ_Q given by equation (6) leads to a more uniform reliability level than the constant value γ_Q = 1,5.

The theoretical models of basic variables indicated in Table 1 are based on recommendations of JCSS (Joint Committee on Structural Safety) [5] and previous studies [6–8].

Table 1: Models of basic variables.

Variable	Symb. X	Distr.	Partial factor γ	Char. val. X_k	The mean μ_X	CoV V_X
Resistance	R	LN	1,15	From (2)	1,25R_k	0,10
Permanent load	G	N	1,35	From (5)	1,0G_k	0,10
Snow load, 50-years maximum	S_{50}	GU	γ_Q	$s_{s,k}$	1,0$s_{s,k}$	0,22
Resistance uncert.	K_R	N	-	-	1,0	0,05
Load effect uncert.	K_E	N	-	-	1,0	0,10

In Table 1 the symbol ”N” denotes normal distribution, “LN” lognormal distribution with the lower bound at the origin and “GU” Gumbel distribution of maximum values. The partial factor γ_Q applied to the characteristic snow load $s_{s,k}$ equals 1,50 or alternatively $1 + \chi$, as indicated in equation (6).

WIT Transactions on Engineering Sciences, Vol 58, © 2007 WIT Press
www.witpress.com, ISSN 1743-3533 (on-line)

4 Results of reliability analysis

Results of the reliability analysis are indicated in Figures 1 and 2. Figure 1 shows the variation of the reliability index β with the load ratio χ for the fundamental load combination applied in equation (2) and two different partial factors γ_Q: the constant factor $\gamma_Q = 1{,}5$ (recommended value) and a variable one $\gamma_Q = 1 + \chi$.

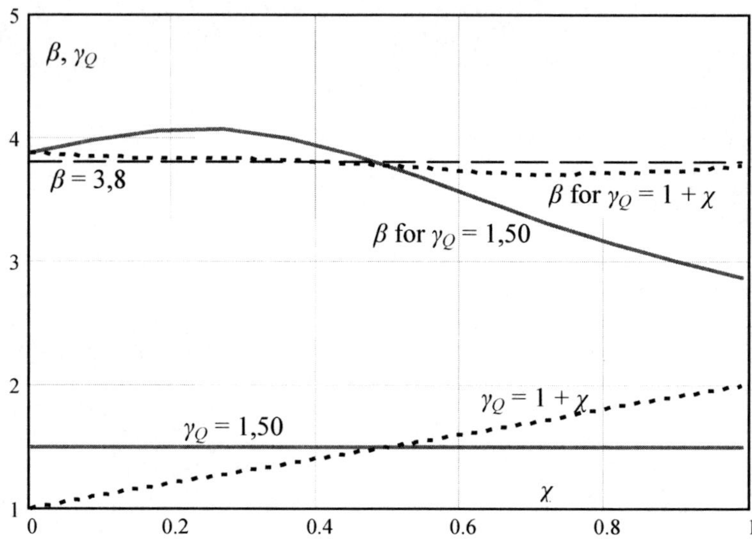

Figure 1: Variation of the reliability index β with the load ratio χ for the partial factors $\gamma_Q = 1{,}5$ and $\gamma_Q = 1 + \chi$.

It follows from Figure 1 that the constant partial factor $\gamma_Q = 1{,}5$ leads to a significant variation of the reliability index with the load ratio χ. Moreover, for the load ratios $\chi > 0{,}5$ the index decreases below the target value of 3,8 recommended in [1] and the reliability of a structural member is insufficient. The partial factor $\gamma_Q = 1{,}5$ seems to be satisfactory for the load ratio $\chi < 0{,}5$ only. For the ratio $\chi > 0{,}5$ the partial factor for snow load γ_Q should be greater than 1,5. Obviously, a significantly more uniform reliability level is obtained for the partial factor $\gamma_Q = 1 + \chi$, which seems to lead to an acceptable reliability level for any load ratio χ.

The results indicated in Figure 1 are confirmed by Figure 2 showing the variation of the reliability index β with the partial factor γ_Q (within the range from 1,2 to 2) for selected load ratios $\chi = 0{,}4$, 0,6 and 0,8. Apparently, to reach the desired reliability index $\beta = 3{,}8$ the partial factor γ_Q should increase with an increasing load ratio χ. For a realistic load ratio $\chi = 0{,}6$ (an expected value for lightweight roofs) the partial factor γ_Q should be about 1,6, thus $\gamma_Q = 1 + \chi$. Similar results are indicated in a recent study [9].

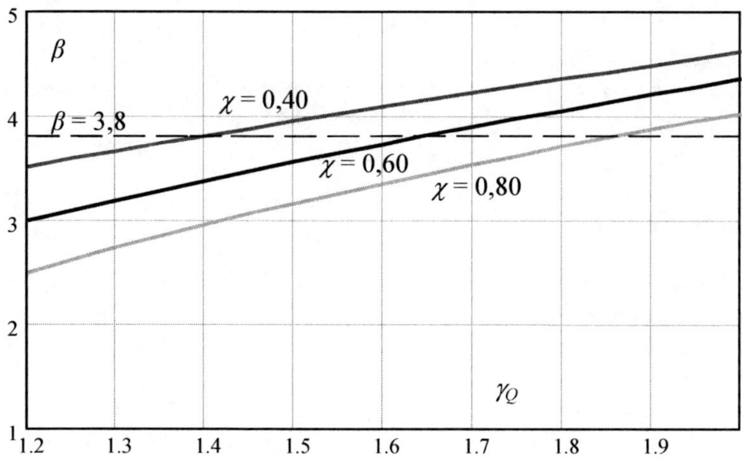

Figure 2: Variation of the reliability index β with the partial factor γ_Q for the selected load ratios χ = 0,4, 0,6 and 0,8.

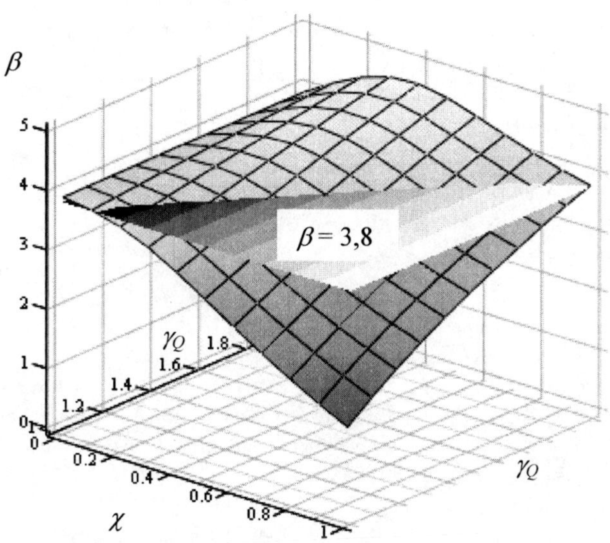

Figure 3: Variation of the reliability index β with the load ratio χ and the partial factor γ_Q.

Figure 3 shows the variation of the reliability index β with both the load ratio χ and the partial factor γ_Q. It can be used for the specification of the adequate partial factor γ_Q for a given load ratio χ, so that the desired reliability level β = 3,8 would be guaranteed. Figure 3 also clearly shows why the partial factor γ_Q should depend on the load ratio χ (as indicated by equation (6)) if a uniform reliability level (a constant index β) is to be achieved.

5 Concluding remarks

The following conclusions may be drawn from the presented study:

1. The constant partial factor $\gamma_Q = 1{,}5$ leads to a significantly variable (non-uniform) reliability level with respect to the load ratio χ.
2. For the load ratios $\chi > 0{,}5$ the index β is less than 3,8 and reliability of a structural member is insufficient.
3. The recommended partial factor $\gamma_Q = 1{,}5$ seems to be satisfactory for a load ratio $\chi < 0{,}5$. For a ratio $\chi > 0{,}5$ the partial factor for snow load γ_Q should be greater than 1,5.
4. To reach the desired reliability index $\beta = 3{,}8$ the partial factor γ_Q should increase with an increasing load ratio χ.
5. A significantly more uniform reliability level is obtained for the partial factor $\gamma_Q = 1 + \chi$, which seems to lead to an acceptable reliability level for any load ratio χ.

It is, however, emphasized that the presented results may be significantly dependent on the assumed models for basic variables and should be considered as informative only.

Acknowledgement

This study is a part of the project GACR 103/06/1521 "Reliability and risk assessment of structures in extreme conditions".

References

[1] EN 1990 Eurocode: Basis of structural design. CEN, 2002.

[2] EN 1991-1-1 Actions on structures – Part 1-1: General actions – Densities, self weight, imposed loads for buildings. CEN, 2002.

[3] EN 1991-1-3 Eurocode 1: Actions on Structures – Part 1-3: General actions – Snow loads. CEN, 2004.

[4] Commission of the European Communities, DGIII-D3, Snow loads, Final Report, University of Pisa, 1999.

[5] JCSS Probabilistic Model Code, http://www.jcss.ethz.ch/, 2005.

[6] Holický M., Marková J. and Sýkora M.: Spolehlivost lehkých střech (Reliability of lightweight roofs), Stavební obzor 2007 (to be published).

[7] Holický, M. - Retief, J.: Reliability Assessment of Alternative Eurocode and South African Load Combination Schemes for Structural Design.

Journal of the South African Institution of Civil Engineering. 2005, vol. 47, no. 1, p. 15-20.

[8] Holický, M. Calibration of load combinations for equal safety of concrete members In: *Proceedings of the* 6th International Congress Global Construction: Ultimate Concrete Opportunities, 5-7 July 2005, Dundee UK, pp. 677-684.

[9] Schleich J.B., Sedlacek G. & Kraus O. Realistic Safety Approach for Steel Structures, In: Eurosteel, Coimbra, p. 1521-1530, 2002.

Safety investigation on rail and road vehicles exposed to cross-wind: wind tunnel tests and multi-body simulations

F. Cheli, R. Corradi, D. Rocchi, E. Sabbioni & G. Tomasini
Mechanical Engineering Department, Politecnico di Milano, Italy

Abstract

A methodology aimed at evaluating the limit safety conditions of road and rail vehicles when exposed to cross wind is presented in this paper. The procedure integrates wind-tunnel experimental tests and numerical simulations with Multi-Body (MB) vehicle models. The main results of the proposed methodology are the vehicle Critical Wind Curves, representing the combination between wind and vehicle speed, corresponding to the overcoming of the vehicle safety limit.
Keywords: MB simulations, wind tunnel testing, safety, aerodynamic forces.

1 Introduction

Over the years the effects of strong winds on road and rail vehicles have become of increasing concern for transportation system operators. The main risk associated with cross wind is the vehicle rollover, which is particularly critical when the vehicle exits a tunnel or when the vehicle runs in a curve on exposed sites such as viaducts, embankments or long-span bridges. Traffic disruption, economic loss, injury, mortality and risks connected with the transported goods, such as hydrocarbons, toxic, flammable and explosive substances, etc. result as a consequence.

Aim of the work is to describe the methodology developed for assessing of the limit safety conditions of road and rail vehicles in terms of Critical Wind Curves (CWC) that represent the combination of vehicle speed and wind mean speed which leads to the overcoming of the safety limits. The methodology integrates wind-tunnel experimental tests and numerical simulations with Multi-Body (MB) vehicle models. The experimental analysis (wind tunnel testing)

WIT Transactions on Engineering Sciences, Vol 58, © 2007 WIT Press
www.witpress.com, ISSN 1743-3533 (on-line)
doi:10.2495/EN070071

allows one to evaluate both the static aerodynamic coefficients, from the measurement of the static component of the aerodynamic forces, and the aerodynamic admittance function. The vehicle transient dynamic response when exposed to cross-winds is then assessed through simulations with MB models of both rail and road vehicles. The unsteady aerodynamic loads due to the turbulent wind acting on the vehicle are calculated through an algorithm, based on both static aerodynamic coefficients and admittance function, measured during wind tunnel testing on reduced scale vehicle models.

2 Numerical experimental methodology

The proposed numerical-experimental methodology for evaluating the Critical Wind Curves of a road/rail vehicle can be divided into two phases:

- experimental measurements performed during wind tunnel tests on scale models;
- numerical analysis to evaluate the dynamic response of a vehicle exposed to turbulent cross wind.

Figure 1 shows the flow chart of the overall numerical-experimental procedure. The inputs to the numerical algorithm for the evaluation of the aerodynamic forces due to the cross wind action on the vehicle are:

1. static aerodynamic coefficients of the vehicle;
2. aerodynamic admittance function;
3. turbulent wind speed distribution as a function of time and space.

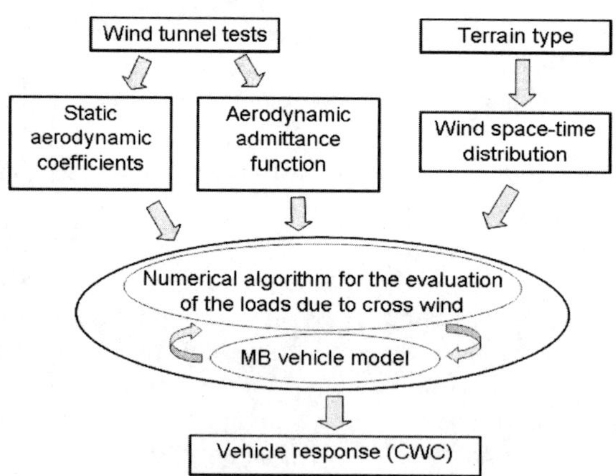

Figure 1: Experimental-numerical methodology flow-chart.

With reference to points 1. and 2., the experimental tests allow to evaluate both the static aerodynamic coefficients, from the measurement of the static component of the aerodynamic forces, and the aerodynamic admittance function. This function accounts for the spatial correlation of pressures at any two points on the vehicle's surface and it represents a modifying adjustment of the ideal case of a vehicle enveloped by turbulent wind with full spatial correlation [1].

About the third point, since turbulent wind is a random process, a time-space distribution of the wind speed can be simulated for a given mean wind speed and terrain type, known the statistical properties of the atmospheric boundary layer: wind speed Power Spectral Density (PSD), turbulence intensity, integral length scales [2]. The numerical algorithm for the evaluation of the aerodynamic loads is directly interfaced with a MB model for the calculation of the vehicle dynamic behaviour (Figure 1). The limit safety conditions are then determined by evaluating the vehicle Critical Wind Curves (CWC).

3 Wind tunnel tests

The first part of the proposed methodology consists in performing wind tunnel tests on scale vehicle models in order to experimentally measure the vehicle mean aerodynamic coefficients and the vehicle aerodynamic admittance function (inputs of the numerical algorithm for the evaluation of the aerodynamic forces, Figure 1). Different kind of vehicles have been tested along the years: ETR500, ETR480, EMUV250 and IC train UIC-Z1 control trailer coach (Figures 7 and 8), for rail vehicles, and lorry, tank, articulated vehicle (Figure 5), for road heavy vehicles.

Figure 2: Politecnico di Milano wind tunnel.

Table 1: Main characteristics of the Politecnico di Milano wind tunnel.

Max. installed power		1.5	[MW]
Global dimensions		50x15x15	[mxmxm]
Test section	Dimensions [m]	Max. vel. [m/s]	Δu/u [%]
Civil	14x4	18	< ± 2
Aeronautical	4x4	60	< ± 0.2

Tests are usually carried out in the civil section (see Table 1) of the Politecnico di Milano wind tunnel (see Figure 2), which allows testing large scale models, together with significant reduction of undesirable edge and blockage effects.

Different flow conditions can be reproduced during the tests in order to evaluate the effect of the wind characteristics on the aerodynamic coefficients. Low turbulence flow is obtained in the standard wind tunnel operating conditions. It is characterised by a uniform vertical profile of the mean wind velocity (see Figure 3). By properly adding roughness elements and spires before the test section (Figure 6), different atmospheric boundary layer can also be reproduced (turbulence intensity up to 30%). As an example, Figure 4 shows the comparison between the wind speed vertical profile experimentally measured during full scale tests and the one reproduced in the wind tunnel within the WEATHER European project [13].

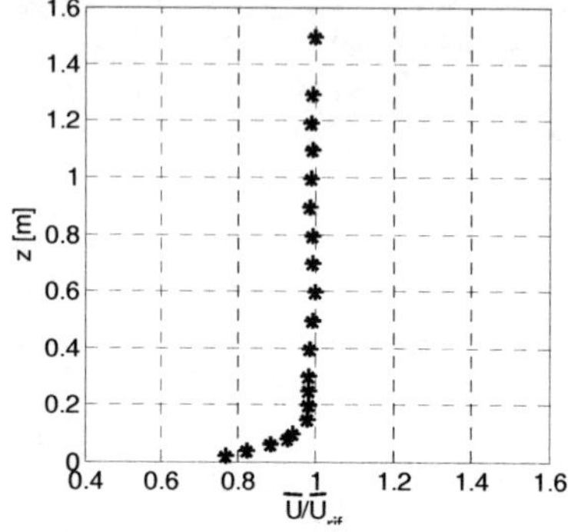

Figure 3: Low turbulence flow: normalized vertical profile of mean wind speed.

Figure 4: High turbulent flow: normalized vertical profile of the mean wind. Comparison with full scale test.

According to the TSI standard, the reference configuration for the static aerodynamic coefficients evaluation is the flat ground (Figure 6a and Figure 7). However, in order to take into account the vehicle real working conditions, several scenarios can be reproduced (viaduct, embankment, presence of solid porous fences, see Figures 5–7). Tests are carried out considering both upwind and downwind conditions.

Figure 5: Tested road heavy vehicles (lorry, tank, articulated vehicle).

The aerodynamic forces and moments in the three directions are measured by means of a six-components dynamometric balance, set under the model and connected to the vehicle in correspondence of the wheels. The wind speed is measured by means of both Pitot tubes for the mean wind speed and a Cobra probe, for the measurement of the statistical characteristics of the turbulent wind (turbulence intensity, wind Power Spectral Density function, integral length scale, [1]). Some preliminary tests have also been carried out with a moving train (tunnel exit, Figure 8), in order to confirm that the aerodynamic coefficients measured in static and in dynamic conditions are the same for the same relative wind speed (hypothesis of the numerical algorithm for the aerodynamic loads evaluation). During these tests, a six components dynamometric balance has been placed between the boogies and the carbody in order to measure the aerodynamic forces and moments acting on the carbody itself.

Once the aerodynamic forces and moments (F_i and M_i) and the mean wind speed $\bar{U}$ have been measured, the corresponding aerodynamic coefficients can be evaluated from the quasi-steady theory [1,4,5]:

$$F_i = \frac{1}{2}\rho_a A C_{Fi}\bar{U}^2; \qquad M_i = \frac{1}{2}\rho_a A h C_{Mi}\bar{U}^2 \qquad (i = x, y, z) \qquad (1)$$

where A stands for the lateral surface of the carbody, h is a reference height and ρ_a is the air density.

(a)

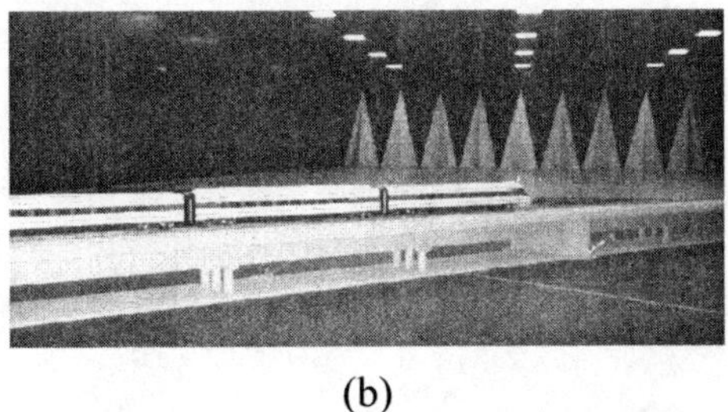

(b)

Figure 6: Boundary layer simulations: (a) Lorry model on flat ground. (b) Train model on viaduct.

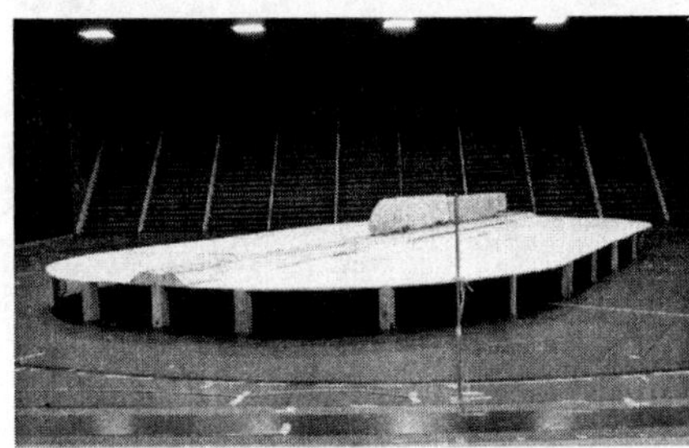

Figure 7: EMUV250 on flat ground.

Figure 8: Dynamic test (ETR470): tunnel exit.

As an example, the influence of the scenario (flat ground and embankment) is analysed in Figure 9 for the ETR500 train. The overturning moment coefficient is shown as a function of the yaw angle α.

The tests carried out in atmospheric boundary layer conditions also allow to identify the vehicle aerodynamic admittance function H [1]: the square of the aerodynamic admittance function is evaluated [4] as the non-dimensional ratio between the PSD of the measured aerodynamic force and the PSD of the aerodynamic force evaluated with the quasi-static theory, starting from the wind speed in the same experimental tests and the static aerodynamic coefficients (eq. (1)). As an example, the ETR500 aerodynamic admittance function is shown in Figure 10 as a function of the adimensional frequency $f\,{}^{x}L_U/\bar{U}$, where f is the frequency and ${}^{x}L_U$ is the longitudinal integral scale.

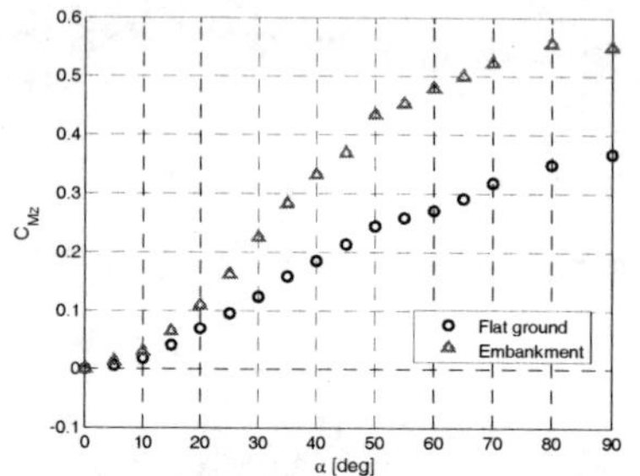

Figure 9: ETR500: overturning moment coefficient, influence of the scenario.

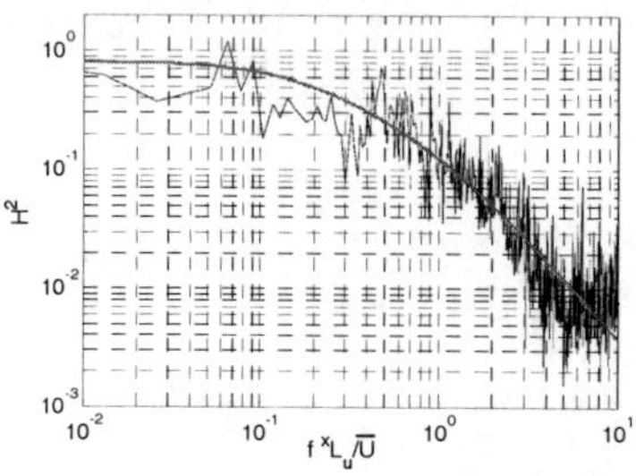

Figure 10: ETR 500: admittance function; experimental data and interpolating function.

4 The numerical algorithm for wind loads evaluation

A numerical algorithm has been set up for the aerodynamic loads evaluation. The adopted approach is based on the corrected quasi-steady theory, a classic approach for many applications [2]. This theory consists in applying the quasi-steady theory (eq. (1)), rigorous for the calculation of the aerodynamic force mean value, and in correcting it, in the frequency domain, by the admittance function [4,5]: thus it represents a modifying adjustment of the ideal case of a vehicle enveloped by turbulent wind with full spatial correlation.

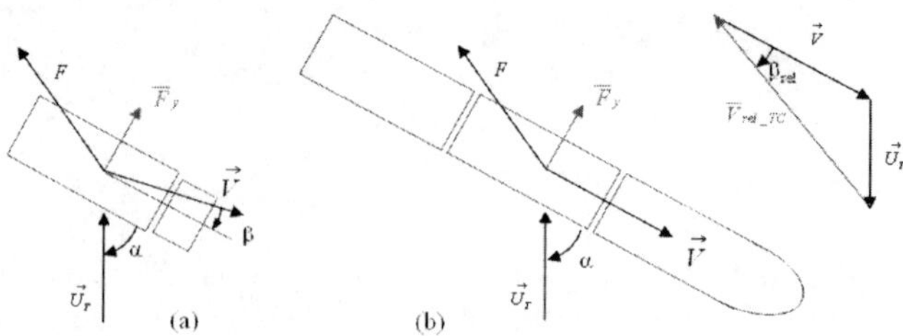

Figure 11: Corrected relative vehicle-wind velocity V_{rel_TC} and related angle of attack β_{rel}: (a) road vehicles, (b) rail vehicles.

It is to point out that, for rail vehicles, being the trajectory imposed, the relative angle of attack (β_{rel},, see Figure 11) is determined once the wind direction (angle α) and the track geometry are known. Thus the time history of the aerodynamic loads (input of the MB vehicle model) can be calculated in a pre-processing phase. Of course this is not possible for road vehicles, since the vehicle path (and consequently the relative angle of attack) depend on the external forces acting on it and on the driver response. In this case, the numerical algorithm for aerodynamic forces calculation is directly interfaced to the MB vehicle model (see Figure 1): during the integration process, at each time step, it gets as an input, from the dynamic model, the vehicle speed and the vehicle sideslip angle β (that allows the evaluation of the vehicle trajectory, see Figure 11) and it gives as an output, to the vehicle model, the instantaneous value of the aerodynamic loads.

5 Vehicle MB models

The vehicle dynamic response to cross wind forces is evaluated through MB rail and road vehicle models developed by the Mechanical Engineering Department of Politecnico di Milano.

5.1 Rail vehicle model

In order to simulate the dynamic response of a rail vehicle subjected to cross wind, the algorithm for the aerodynamic loads calculation, has been implemented in A.D.Tre.S., a rail vehicle dynamic simulation program [3], developed by the Mechanical Engineering Department of Politecnico di Milano. A.D.Tre.S. is suitable for simulating the motion of a railway vehicle running on tangent track and curve, also taking into account the deformation of track and substructure and the rail and wheels surface corrugation. The vehicle model is composed of a car body, two bogies and four wheelsets, introduced as rigid or flexible bodies, by means of modal superposition approach [3]. Car body, bogies and wheelsets are linked one to each other by means of the elastic and damping elements reproducing the primary and secondary suspensions (see Figure 12). The simulation code accounts for the motion of a train-track system in both vertical and lateral direction, while the longitudinal motion of the vehicle is assumed to take place with constant velocity. The vehicle equations of motion are written with respect to a moving reference system, travelling with constant speed along the ideal track centreline.

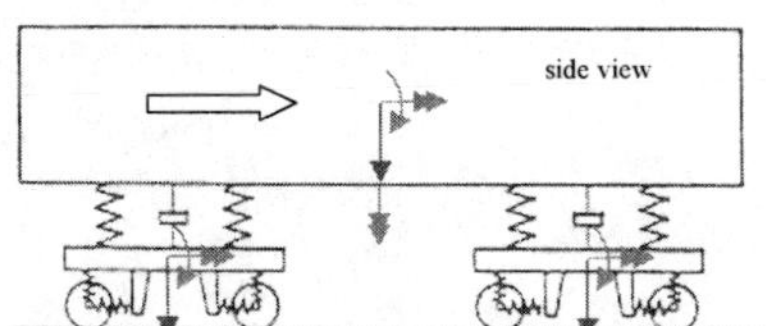

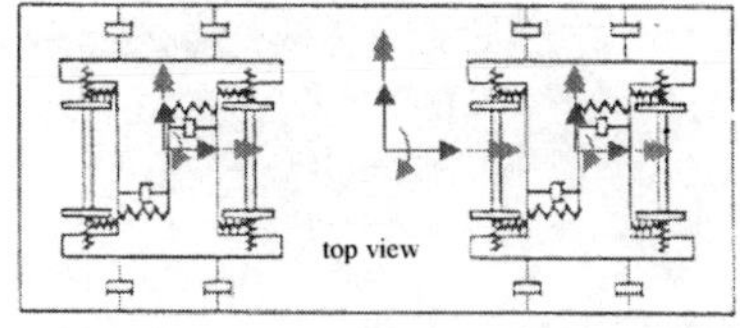

Figure 12: The rail vehicle model.

The contact model is based on a preliminary geometrical analysis on the measured wheel and rail profiles. Contact parameters such as rolling radius and contact angle are reported in table form, as functions of wheel-rail lateral displacement. The geometrical analysis allows also to determine the number of the potential contact points for a given wheel-rail lateral displacement. Lateral and longitudinal creepages are computed and tangential and longitudinal forces at each active contact are obtained according to Shen, Hedrick and Elkins formulation [3].

The equations of motion are integrated in time domain, by means of a numerical step by step procedure [3]. In the present analysis, the simplifying assumptions of "infinitely rigid" track and vehicle rigid bodies were made.

5.2 Road vehicle model

In the case of road vehicle dynamic simulation, the algorithm for wind loads evaluation has been interfaced with a 14 d.o.f. vehicle model developed by the Mechanical Engineering Department of Politecnico di Milano [9,10]. The vehicle is considered as composed by five rigid bodies: the vehicle chassis (the sprung mass), having 6 d.o.f., and the four unsprung masses (the four wheels) each characterised by wheel rotation and a vertical movement. The vehicle chassis and the unsprung masses are connected through the suspension system, while the unsprung masses are linked to the ground by means of a spring-damper element, representing the tires radial stiffness and damping (see Figure 13). The suspensions have been introduced through multi-dimensional tables in order to take into account their elasto-kinematic behaviour. The engine is introduced through its characteristic curve, i.e. driving torque versus angular speed, while the deformability of the driveline is modelled through a first order system. The tires behaviour is described using MF-Tyre model version 2002 [11].

Steer angle, brake pressure, throttle valve position and gear are all regarded as imposed inputs characterising a particular manoeuvre and they can be either introduced by the user in pre-processing phase (open loop manoeuvres) or computed during the simulation by a driver model (closed loop manoeuvres).

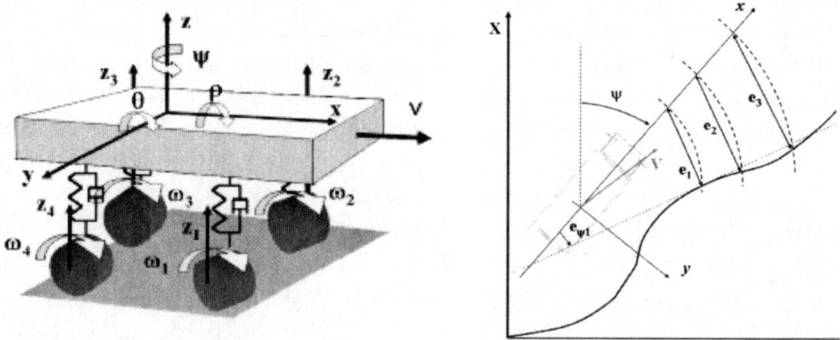

Figure 13: The 14 d.o.f. road vehicle model.

Figure 14: The road vehicle driver model.

In order to obtain comparable results (wind forces acting on the vehicle depending both on the vehicle trajectory and on the wind speed time-space distribution, section 4), a driver model has been implemented to complete the simulation tool. The steering angle is determined through a PID (Proportional-Integrative-Derivative) regulator acting on the error between the actual and the desired vehicle trajectory at three preview points (e_1, e_2, e_3 in Figure 14). In the steering angle calculation the nearest preview point is the most weighted.

6 Simulation results

The developed simulation tool consisting of the algorithm for the aerodynamic loads calculation and the MB vehicle model, can be used to evaluate the safety limit of the vehicle for each running condition. The index used to determine the Critical Wind Curves is the overturning coefficient (the Prudhomme criterion or the Nadal's coefficient can be considered as alternative indexes for defining the CWCs for rail vehicles [4]), defined as:

$$\eta = \frac{\sum_i \left| Q_{left} - Q_{right} \right|}{\sum_i \left(Q_{left} + Q_{right} \right)} < \eta_{\lim} = 0.9 \qquad \left(i = bogie\, or\, axle \right) \tag{2}$$

where Q are the vertical loads acting on the wheels of each boogie/axle.

By comparing the time histories of the index with the corresponding limit value, it is possible to find whether the combination of vehicle speed V and gust wind speed U_{lim} corresponds or not to the overcoming of the safety limit for the considered vehicle: the critical wind curves (CWC) represent the gust wind speed U_{lim} that leads to the overcoming of the safety index as a function of the vehicle speed V.

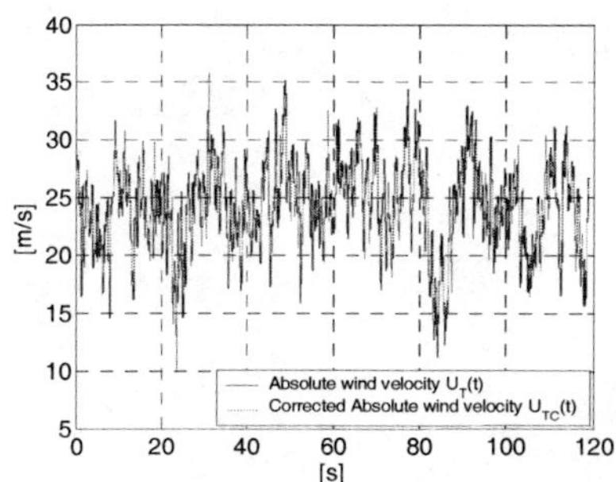

Figure 15: Tangent track simulation: time history of the wind speed, I_U=19%, ^xL_U=90m (input of the MB vehicle model).

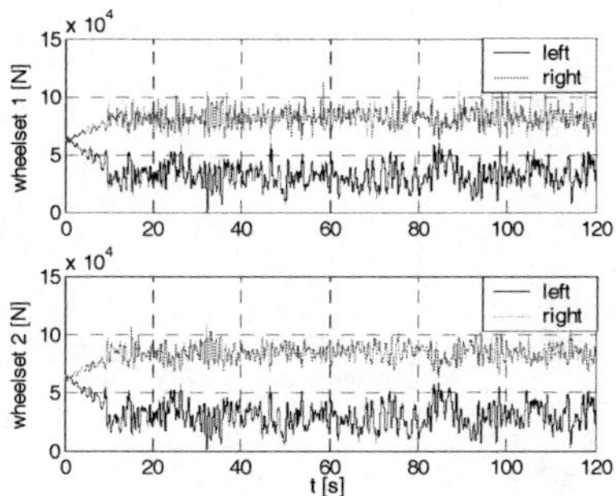

Figure 16: Tangent track simulation: time history of the vertical loads, I_U=19%, ^xL_U=90m (output of the MB vehicle model).

Figures 15 and 16 show an example of input and output time histories. A rail vehicle running in a tangent track and it is subjected to a turbulent wind having the following characteristics: $\bar{U}$=25m/s, turbulence intensity I_u=19% and integral length scale ^xL_u=90m. More in particular, Figure 15 shows the cross wind speed time history (input of the MB vehicle model), while in Figure 16 the time histories of the wheel vertical loads are reported (output of the MB model). Since the wind is coming from the left, a load transfer occurs from the left to the right wheels.

Starting from the simulation results, the safety index and the CWCs can thus be evaluated. Since turbulent wind is a random process, a statistical analysis can be performed: the resulting CWC can be evaluated as the mean value of several simulations carried out for different random generations of the turbulent wind speed given as an input to the MB vehicle model. As an example, Figure 17 shows the mean CWC and the relative dispersion band obtained for a train on flat track during a tangent track simulation (blue triangles). The mean CWC obtained through the proposed methodology (blue triangles) is compared with the one obtained using the TSI methodology (red crosses). As it can be seen, the TSI CWC is within the dispersion band.

The influence of several overall parameters can be taken into account and evaluated through the proposed methodology: the deformability of brides [12], the wind turbulence intensity, the considered scenario (flat ground, viaduct embankment, [4,5]), the track/road irregularity [4], etc. In Figure 18, the influence of the scenario during a tangent track simulation with a lorry is evaluated. As can be noticed, the flat ground is always the less critical situation. At low vehicle speed the embankment and the viaduct present the same trend, while at high vehicle speed the embankment seems to be the most critical condition. This is due to the fact that the wind flow is accelerated by the trapezoidal cross-section of the embankment.

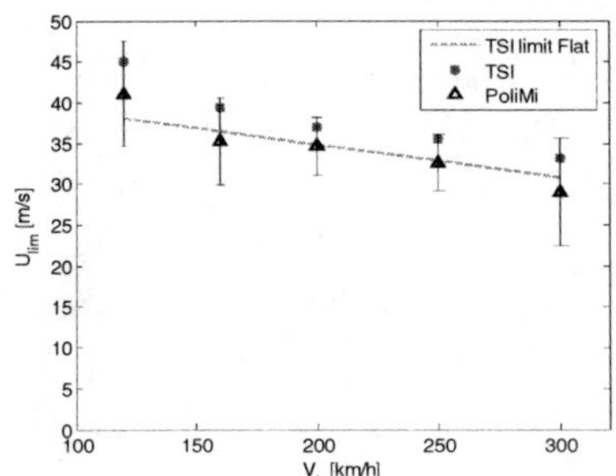

Figure 17: CWC: comparison between the PoliMi and TSI methodology, for a train, flat ground, tangent track.

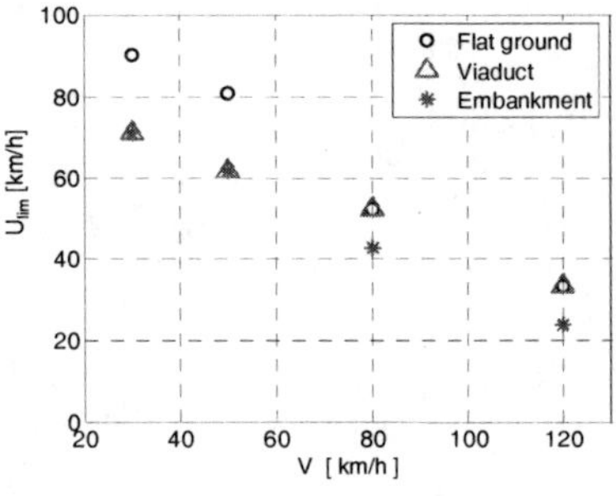

Figure 18: CWC: influence of the scenario. Lorry, tangent track.

7 Concluding remarks

In the present paper a numerical-experimental methodology for the evaluation of vehicle limit safety conditions when exposed to cross wind has been presented, applicable to both rail and road vehicles. Wind tunnel experimental tests are needed in order to evaluate the vehicle aerodynamic coefficients in different operating conditions and to identify the vehicle admittance function. These are the inputs to the numerical algorithm for the turbulent wind loads evaluation (based on the corrected quasi-static theory), which is directly interfaced with MB road/rail vehicle models. Through this simulation tool it is thus possible to assess the vehicle dynamics when exposed to cross wind and finally evaluate the safety limit conditions in terms of Critical Wind Curves. The proposed methodology has proved to be a useful tool for the design of on board vehicle controllers and infrastructure alert systems.

References

[1] Simiu, E., Scanlan, R.H., 1986, *Wind effects on structures*, New York, Wiley-Interscience.

[2] Cheli, F., Curami, A., Bocciolone, M., Zasso, A., 1991, *Wind measurements on the Humber Bridge and numerical simulation*, 8-th Int. Conference on Wind Engineering ICWE, London, Canada.

[3] Bruni, S., Collina, A., Diana, G. and Vanolo, P.: *Lateral Dynamics Of A Railway Vehicle In Tangent Track And Curve: Tests And Simulations*. Proceedings of 16-th IAVSD Symposium, Vol. 33 pp 464-477 (1999).

[4] Cheli F., Corradi R., Diana G., Tomasini G., *A Numerical-Experimental Approach to Evaluate the Aerodynamic Effects on Rail Vehicle Dynamics*, 18th IAVSD Symposium August 24 to 30, 2003.

[5] F. Cheli, P. Belforte, S. Melzi, E. Sabbioni, G. Tomasini: *A numerical-experimental approach for evaluating cross wind aerodynamic effects on heavy vehicles*, Proc. of XIX IAVSD Symposium, Milano, Italy, 29 August-2 September 2005.

[6] Bocciolone, M., Cheli, F., Corradi, R., Diana, G. and Tomasini, G.: *Wind Tunnel Tests For The Identification Of The Aerodynamic Forces On Rail Vehicles.* Proceedings of 11th ICWE – International Conference on Wind Engineering, June 2003.

[7] S. Bruni, F. Cheli, E. Sabbioni, S. Sciacca, G. Tomasini, *Studio sperimentale delle caratteristiche aerodinamiche di veicoli stradali pesanti*, IX Convegno IN-VENTO, Pescara, Italy, 19-22 June 2006.

[8] Cooper R. (1984), *Atmospheric Turbulence with respect to moving ground vehicles*, Journal of wind engineering and industrial aerodynamics, vol. 17, 215-238

[9] F. Cheli, E. Leo, S. Melzi, F. Mancosu, E. Giangiulio, D. Arosio, 2006, Implementation of a 14 d.o.f. model for the prediction of vehicle dynamics and is interaction with active control systems, Tire Technology Expo, Stutgart, Germany, 2006.

[10] F. Cheli, E. Leo, S. Melzi, F. Mancosu, A 14 d.o.f. model for evaluation of vehicle's dynamics, International Journal of Mechanics and Control, Vol. 6, No. 2, pp. 19-30, 2005.
[11] H. B. Pacejka, Tyre and Vehicle Dynamics, Butterworth and Heinemann Editions, 2002.
[12] F. Cheli, A. Collina, E. Leo, F. Resta, G. Tomasini, Numerical-experimental methodology for runnability analysis and wind-bridge-vehicle interaction study, III ECCM Conference, Lisbon, Portugal, June 2006.
[13] D. Delaunay, C. J. Baker, F. Cheli, H. Morvan et al., Development of wind alarm systems for road and rail vehicles: presentation of the WEATHER project, 13th Int. Road Weather Conference, 25th -27th March 2006.

Sustainable management of Prespa Lake

V. Popov[1], T. Anovski[2] & R. Gospavic[1]
[1]*Wessex Institute of Technology, Environmental and Fluid Mechanics, Southampton, UK*
[2]*Faculty of Technology and Metallurgy, Department of Chemical and Control Engineering, Skopje, Republic of Macedonia*

Abstract

The Prespa Lake is on the border between Greece, Albania and Republic of Macedonia. The importance of Prespa Lake has been widely recognised by national and international bodies in the three neighbouring countries not only because of its natural beauty, but also because of its high biodiversity, including populations of rare water birds, like for example the Dalmatian pelican, as well as for its cultural values including Byzantine monuments. In recent years a decrease in water in the lake has been recorded. Though there have been several studies investigating the water level decrease, it has been impossible so far to conclude on the reasons behind this phenomenon. A research project has started recently supported by the NATO SfP programme which has as one of the main objectives to look into the cause behind the water level decrease, and also to suggest strategies for the water usage of the lake's watershed.

1 Introduction

Three lakes, Ohrid, Big Prespa and Small Prespa, are on the borders between Albania, Republic of Macedonia and Greece, see Figure 1. Galichica and Dry mountains separate the lakes. According to an existing hypothesis by Cvijic [1], water from the Prespa Lake, which is shared by the three neighbouring countries, is drained through the Galichica and Dry mountains into Ohrid Lake. In Figure 2 the lakes and the Galichica and Dry mountains are shown. Field research conducted first by Anovski *et al.* [2] and later by other scientists confirmed this hypothesis.

WIT Transactions on Engineering Sciences, Vol 58, © 2007 WIT Press
www.witpress.com, ISSN 1743-3533 (on-line)
doi:10.2495/EN070081

The Lakes Big Prespa (253.6 km^2) and Small Prespa (47.4 km^2) are at 850 m asl. and are linked by a small channel with a sluice that separates the two lakes, see Figure 1. In the past, periodical oscillations of the lakes' level were in the range of one to three metres, depending on the amount of rain in the season. After the mid 80's, a steady decrease of the water level has been recorded that disturbs the ecological balance of the lake and the watershed area resulting in serious consequences for the fishing and tourist industry in the trans-boundary Prespa region. In addition to this, the industrial activities as well as the overuse of the herbicides in agriculture activities raised the problem of pollution of the water in the Prespa Lake.

Figure 1: Ohrid, Big Prespa and Small Prespa lakes.

The Prespa Lake that is examined in the proposed programme constitutes an area of high ecological importance, because of its beauty as recreational places. On the other hand this is a site of economic significance, including agricultural, industrial and other activities, which use the water potential of this natural resource. For all of the above factors the sustainable development of the lakes and their surroundings is a field of particular interest.

The importance of Prespa Lake has been widely recognised by national and international bodies not only because of its natural beauty, but also because of its high biodiversity, including populations of rare water birds, like for example the Dalmatian pelican, a world vulnerable species, as well as for its cultural values including Byzantine monuments.

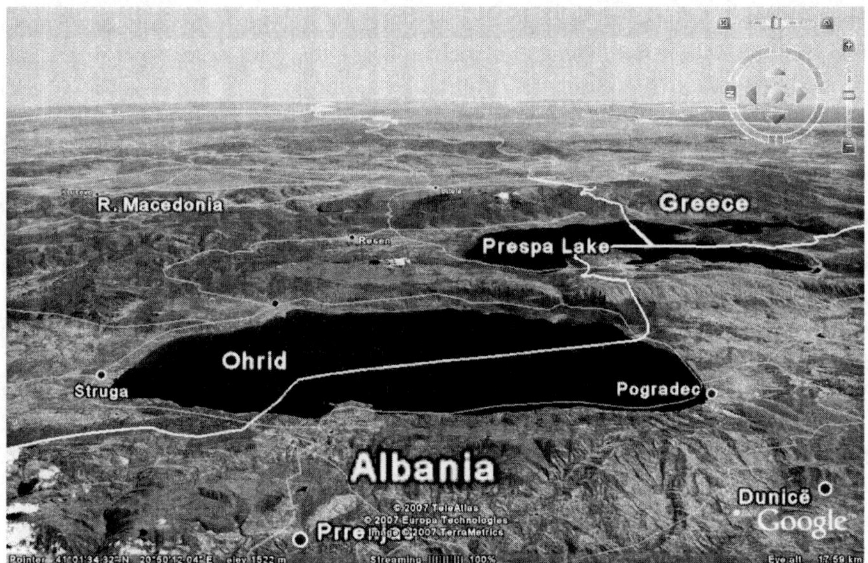

Figure 2: Ohrid and Prespa lakes with the Galichica and Dry Mountains in between.

State authorities of the three countries have enforced the protection status of Prespa through the use of national and international legislative means. A large part of the lakes and their catchment basin has been characterized as a National Park (Albania and Greece) or/and a Wetland of International Importance under the Ramsar Convention (Greece, R. Macedonia).

The willingness of the three governments to co-operate in order to promote the protection of all the Prespa area was corroborated on the 2nd February 2000, when the Prime Ministers of Albania, Greece and the R. Macedonia issued a trilateral declaration recognising the international importance of the Prespa Lakes as well as the need for co-operation in order to promote conservation of its natural and cultural values.

One of the major environmental issues today that may, as well, have implications on the relation between the three neighbouring countries, is the question of global environmental degradation and increased shortages of renewable resources, especially international waters and watercourses, joint river basins and joint lakes. Increased pressure on the environment and natural recourses threatens the economic potentials of the countries. Hence, common response and action in the preservation of natural resources should be on the top of the agendas of all different regional initiatives and endeavors among the countries in the region. This calls for the development of instruments for managing of transboundary surface water and groundwater in the framework of ECE Helsinki Convention on the protection and Use of Transboundary Waters and International Lakes. This involves the collaboration for a regional characterisation of quantity and quality assessment and controls of water use and pollution emissions.

The largest fraction of the population which live in the Prespa Lake's area depend on the lake itself. The lakes offer natural beauty that used to attract many tourists every year and they also offer suitable conditions for fish industry and agriculture. If the ecosystem of the lakes is disturbed, and there are indications that this is becoming reality, this may have disastrous consequence on the lives of a large part of the population in the area forcing many of them to migrate to other regions, which would create problems for those areas.

A joint project between five institutions from R. Macedonia, Albania, Greece and UK is currently funded through NATO Science for Peace and Security programme. The main objectives of the project are: (i) to understand the mechanism behind the water loss in the Prespa lake; (ii) to prepare recommendations for sustainable use of the water in the Prespa Lake watershed.

It is impossible to quantify the importance of this problem, since it involves the future of a natural resource of unique beauty and the future of the local communities whose economy largely depend on the lake.

2 Results available from previous studies

Recent research programmes that included research teams from the three neighbouring countries have established that the water level in the Big Prespa Lake has been lowering and the connection with the Small Prespa Lake has decreased (actually it is currently interrupted).

2.1 Geology

A great variety of rocks concerning their age, genesis, and lithology constitute this area. Intensive tectonic implications result in the modification of the primary rocks and in the formation of very different structures and of a heterogeneous relief picture. The following three main rock complexes are identified in the study area: metamorphic and intrusive rocks, carbonate rocks and terrigene rocks.

2.2 Hydrogeology

From the hydro-geological point of view, the rocks of Prespa-Ohrid area are classified as porous aquifers, karstic and fissured aquifers, fissured aquifers and practically non-aquiferous rocks.

2.3 Quality of water

Recent studies included water temperature, dissolved oxygen content, pH, pesticides (HCB, PCB 138, γ-HCH, DDT, etc), different chemicals (Ca, Mg, Na, K, Fe, Cl, HCO_3, NO_3, etc), and radioactivity (total-α, total-β and γ-spectrometry) sampled at different points periodically for the duration of the research programme. The comparison of the recent data with the ones of the past years shows that there is a significant deterioration of the water quality.

2.4 Isotope data

There is significant amount of data, collected by teams from the three neighbouring countries, on the $d^{18}O$ dating back to 1980. This data can be used to estimate the retention time, to calibrate the numerical models for water movement in the Lake area and to trace any changes in the circulation of the water in the Lake.

2.5 Hydrology and water balance

There has been an exchange of meteorological data between the three neighbouring countries and cooperation between professional institutions in order to define the water balance of the Prespa Lake. The previous studies on water balance included: precipitation, evaporation, inflow and outflow by surface or underground mechanism in each of the three countries. These studies will be used as reference points for the future water balance estimations. However, the data from the three countries was never complete during a whole year, and this is the main reason that there was a discrepancy of $\approx 10^8$ m^3 per year. The measurements during the proposed research programme will help a more accurate estimate to be obtained.

2.6 Charting and profiling of the bottom of the Lake

During the years 2000 and 2002 charting and profiling of the bottom of the Big Prespa Lake was carried out. The results showed two characteristic trenches-like structures on the lake's bed. One of the trenches is 7 km long, 0.9 km wide and 35 m deep on average, while the other one is 12 km long, 1.5 km wide and 23 m deep on average. The south part of the Lake showed an unexpected structure of the lake's bottom with sharp faults indicating strong tectonic movements. In the eastern part of the lake a large sedimentation area was identified, where several rivers inflow the lake. However, there is still plenty of data missing, and there is also some inconsistency, like using different equipment with different methodology in different parts.

2.7 Artificial tracer experiment

A preliminary artificial tracer experiment was carried out in order to obtain information on the dynamical parameters of the big Prespa Lake's water movement through the Galichica karstic massif. The experiment confirmed that an underground hydraulic connection between the Prespa and Ohrid lakes exists. At some observation points there was only one while at other points there were several peaks in the concentration of the observed tracer. This shows the complexity of the karstic system separating the two lakes. The results of this experiment will be used in order to improve the design of the next artificial tracer experiment, especially the sampling regime, which was not optimal during the first experiment and therefore the interpretation of the results was difficult.

3 The problem that will be addressed in the Project

The main problem, which will be addressed in the NATO SfP Project is the decrease of the water level in the Prespa Lake and the mechanism behind it. In order an accurate water balance to be performed, measurements of precipitation, evaporation and direct river inflow into the lake during certain period of time, e.g., one year, must be conducted in the three neighbouring countries. Though each country performs such measurements, they were not systematically performed for a whole year in the three countries. This effectively prevents an accurate estimate of the water balance of the Prespa Lake.

In order that the dynamic parameters of the hydraulic communication between Prespa and Ohrid lakes through Dry Mountain and Galichica are established, tracer experiment will be conducted, which will include injection of tracers into sinkholes in the Prespa Lake and monitoring in the springs of the Ohrid Lake.

The following economic and social impacts in the region have been observed:

(i) 44 km of beaches on the Macedonian (FYRO) side have been disturbed because the Lake's shoreline falls and leaves muddy ground. Such situation is expensive to remediate since the shoreline is changing. There are approximately 7000 beds on the Macedonian part of Prespa Lake. In the past there were approximately 250,000 to 300,000 visitor/nights in the hotels per year, while in the past 7 to 8 years this number is at most 30,000 per year. The main reason for the decrease of the number of tourists is the decrease of the water level of the Lake.

(ii) As the water level in the Lake decreases, the water temperature in the Lake is changing as well. This affects the flora and fauna in the lake. The decrease in the fish production affects the fish industry.

(iii) Only about half of the wastewater released, produced by approximately 17,000 inhabitants on the Macedonian side, is pre-treated. As the amount of wastewater released remains more or less constant, except during the summer period when it is higher because of the visitors in the area, and as the water amount in the Lake decreases, the water quality deteriorates.

(iv) The Lake's water is used for agricultural purposes in all three neighbouring countries. It is estimated that within the Prespa Lake region there are 6,500 ha of agricultural land. On one hand the decrease of the water level increases the costs of irrigation since nowadays two pumps in line must be used, because of the lower level of the water, in order to provide the water for the agricultural fields. On the other hand, the water in the Lake during dry seasons may be decreasing by 1 cm per day due to combined effects of evaporation and irrigation. Many of the agricultural fields are placed in the catchment area of the Lake, which contributes that pesticides and fertilizers pollute the Lake. Since the intensity of the source of pollution is more or less constant and the amount of water decreases, the concentrations of polluting compounds in the Lake increase.

(v) The visual beauty of the Prespa Lake suffers due to the decreasing level of the water.

Overall, the cumulative effects of the above-mentioned factors contribute towards ecologic and economic catastrophe of the Prespa Lake region.

Once the mechanisms behind the water level decrease are clearer, recommendations will be given of how to use the resources of the Lake in a more sustainable way.

4 Possible causes for the decrease of the water level

In Figure 3 the water level variation in the Prespa Lake is shown in the period 1956 to 2004. It can be seen that the water level decrease in respect to the highest level is approximately 8 m. For a lake with an average depth of 15 m this can have a significant impact.

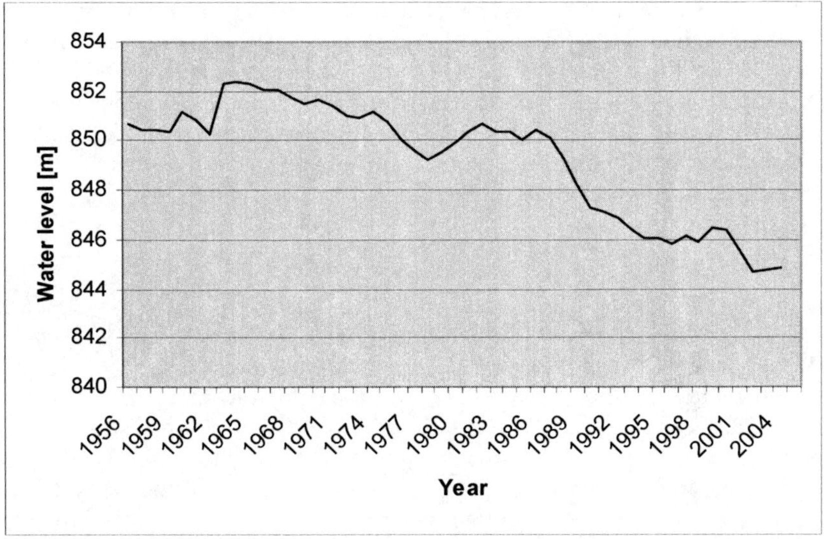

Figure 3: Change of water level of Big Prespa Lake from 1956 to 2004.

The possible causes for the water level decrease have been stated as [3]: (i) tectonic falling of the lake bottom; (ii) widening of underground channels connecting the two lakes; (iii) influence of meteorological parameters.

It is more likely that either (ii) or (iii) is the cause of the water level decrease in the lake, though a combination of (ii) and (iii) cannot be excluded. In the NATO SfP study another cause will be investigated, and that is the anthropogenic factor. Since there are agricultural activities in the area, this can also affect the water level in the lake, especially when there is a dry season. During a dry season more of the water from the lake's watershed, aquifer or direct extraction from the lake, will be used for agricultural purposes, exaggerating in this way the effects of a dry season on the lake's water level.

5 The research programme of the NATO SfP project

The modeling part of the project will provide a link between the results acquired in the past and for the duration of the project. Though in the recent years a joint research programme between the three border countries has been conducted, the cause for this water-level drop is still not known, mainly because the data collected has not been processed in a systematic way and some of the state-of-the-art computer simulation tools have not been used in the previous studies. The modeling part will help calculate the Lake's water budget, based on the surface river inflow, Lake water level, and meteorological data. In order to better understand the hydrological link between the Ohrid and Prespa lakes, some of the new developments in the field of flow in fractured porous media will be used [4].

The novel aspects of the project relate to the way to deal with water resources in karstic regions under the influence of the changing weather patterns. A full evaluation of the data available in the previous years combined with the data collected for the project duration will be used to estimate the implications on a complex system like the Prespa Lake where parameters like sinkholes, precipitation and evaporation, water inflow through rivers and streams and water use in agriculture must be taken into account.

6 Conclusions

The lakes Ohrid and Prespa (big and small) represent a unique and very complex water system, where the water from the Prespa lake leaks into the Ohrid lake through underground pathways. The recent steady water level decrease in the Prespa lake suggests that this beautiful lake which is part of this unique water system can be lost soon, unless the reason behind the water level drop is investigated and efficient measures are proposed for use of the water from the lake's watershed. One of the current efforts in saving the lake is the NATO SfP project, which has as its main objectives to discover the mechanism behind the water level decrease and to suggest strategies for the water usage of the lake's watershed.

It is speculated that the water level decrease could be due to: (i) changes in the hydrological link between the Ohrid and Prespa lakes (increase in the hydraulic conductivity of the aquifer linking the two lakes); (ii) change in weather conditions in recent years; (iii) change in water use by the local population for agricultural and industrial purposes. It is also possible that the cause for the water level decrease is due to combination of the above three possible causes.

The recently started NATO SfP project will include hydrometeorological measurements in the lake's watershed during one year in the three neighbouring countries, in order to accumulate enough accurate data to perform a comprehensive water balance, which should finally distinguish between the three possible causes mentioned above. Adequate models will be used in order to utilize the past and project's data in the most efficient manner. In addition an artificial tracer experiment will be conducted in order to better understand the dynamics of the hydrological link between the two lakes.

Acknowledgement

This work is supported by the NATO Science for Peace and Security programme (Project No. SfP 981116).

References

[1] Cvijic, J. (1906) Fundamentals of Geography and Geology of Macedonia and Serbia, Special Edition VIII+680, Belgrade.
[2] Anovski, T. et al. (1980), "A study of the origin of water in the St. Naum's Springs, Lake Ohrid", FIZIKA – A Journal of Experimental and Theoretical Physics, Vol. 12 (S2).
[3] Anovski, T. (Ed.) (2001) Progress in study of Prespa Lake using nuclear and related techniques. IAEA Regional Project RER/8/008, Faculty of Technology and Metallurgy, Skopje, R. Macedonia.
[4] Peratta, A., Popov, V., (2006) 'A new scheme for numerical modelling of flow and transport processes in 3D fractured porous media', *Advances in Water Resources*, **29**, 42-61.
[5] Samardzioska, T., **Popov**, V. (2005), 'Numerical comparison of the equivalent continuum, non-homogeneous and dual porosity models for flow and transport in fractured porous media', *Advances in Water Resources*, 28, 235-255.

Fire risk assessment of historical areas: the case of Montemor-o-Velho

M. L. A. Santana[1], J. P. Rodrigues[1], A. Leça Coelho[2] & G. L. Charreau[3]
[1]*Departamento de Engenharia Civil, Faculdade de Ciências e Tecnologia, Universidade de Coimbra, Portugal*
[2]*Laboratório Nacional de Engenharia Civil, Lisboa, Portugal*
[3]*Instituto Nacional de Tecnologia Industrial, Buenos Aires, Argentina*

Abstract

Historic constructions are different from contemporary ones for reasons associated with the old conception of villages or towns, and with the types of materials and constructive solutions used. Those characteristics make it difficult to adopt common fire risk assessment methods in the old areas of towns. In Portugal, fire risk analyses became more important after the 1988 fire in Chiado, in the historical area of Lisbon. The features of the area and the absence of any fire detection and suppression facilities led to the destruction of 18 buildings. This fire motivated the development of new fire safety regulations and more careful fire risk analysis countrywide.

This paper first describes the characteristics of the historical areas of Portuguese towns/cities in terms of their buildings, streets, the existing means of first intervention, the characteristics of the fire brigades and other aspects.

Then the methods of fire risk assessment normally used in these areas are summarized and compared, showing their strengths and weaknesses.

Finally, the fire risk analyses carried out on Montemor-o-Velho, an old town in the centre of Portugal, are presented. The different methods used, the results obtained and the fire protection solutions adopted for the different areas and buildings are described.

Keywords: fire, risk, evaluation, heritage, protection.

WIT Transactions on Engineering Sciences, Vol 58, © 2007 WIT Press
www.witpress.com, ISSN 1743-3533 (on-line)
doi:10.2495/EN070091

1 Characteristics of historic areas of Portuguese towns/cities

Historic areas in contemporary urban centres may be important for many reasons: cultural heritage, architecture styles, and special sites within them or social histories. But they may well contain buildings without any special features and even contemporary constructions. The definition of town and city has changed over history, along with their characteristics and concepts. This makes historic zones quite different from contemporary ones in many ways. The streets, the ways of constructing buildings and the types of buildings, society's needs have all evolved, and all present problems that need to be understood and addressed in this paper. The problems may be different in these areas, and common solutions must be adapted before they can be used and in some cases completely new solutions must be created.

How this can be done is described in Câmara Municipal de Guimarães [1], Coelho [2] and Gonçalves [3], with special emphasis on fire safety. The historic areas of Portuguese towns and cities have a number of problems to be solved, or at least to be thought about:

- Narrow, winding streets raise difficulties for vehicles, particularly for Fire Department vehicles carrying the equipment and the water to help extinguish the fire (fig. 1);
- Adjacent buildings with shared separating walls. Fire can thus start and propagate easily through the contact between roofs (fig. 1);
- Buildings in historic areas are residential or small businesses, such as grocery and craft shops. Dwellings are usually too small to justify some measures, but are exposed to fire risk in the same way. Shops often had another use originally, and no changes were made to adjust them to fire safety needs;

Figure 1: Narrow, winding streets, facades face to face, adjacent roofs at the same height, decrepit and contemporary buildings in the same neighbourhood. (Personal Archive.)

- Some buildings have been abandoned or/and are in a poor state of repair with combustible materials inside like wood and old furniture, and there are no plans to restore them. The chance of fire starting accidentally is quite low and there are no people to evacuate, but on the other hand, if a fire spreads to these buildings the material and dust inside can increase the fire.
- Load-bearing structures and facades with very high fire resistance, but inside the horizontal and vertical separation is combustible, which means that in a fire all the building will be exposed to it and it will not be possible to confine the fire just to one or a few compartments;
- There are too many more or less abandoned attics and basements whose access is intricate. Fire can begin very quickly and people will take time to notice it, reach the site and start to tackle the blaze;
- The buildings often have large openings. In the event of a fire with big flames, it will propagate to the upper floors via the windows and also by radiation to other buildings in front of the windows;
- Buildings' electric and gas systems often are old and have not been properly maintained. Short circuits and gas leaks may cause fires.
- As a rule there are not enough fire hydrants to tackle fires, or else they are in disrepair and the water supply sometimes has insufficient pressure;
- There are often no fire alarm or detection systems in these zones, which delays calling the fire brigade and tackling the blaze;
- The population of these areas is mostly elderly, and as such they are not prepared to react rapidly to any type of danger, including fire.
- People have no idea how to store and look after combustible materials in places without adequate access that are not cleaned. Dust encourages the spread of fire.

2 Fire risk assessment in historic areas

Risk analysis is developing all the time, especially in the context of associating high and new technology with mathematics. Risk Analysis is used in many areas and has a real potential for development.

Risk concerns the probable loss in a determined, undesired situation, the likelihood of the situation occurring and all its consequences. Fire risk analysis should therefore focus on the minimization of the risk of fire breaking out, and, if a fire does occur, the minimization of its consequences. It should consider the rapid evacuation of people from dangerous places and extinguishing the fire to prevent it propagating to other rooms and other buildings.

Although big and destructive fires dot the landscape of history fire risk analysis is particularly recent. Even so, there are now several approaches to fire risk analysis of buildings depending largely on the used methods. Towns'

historic centres are obviously their oldest zones, but fire safety measures and analysis have not been applied to them, and it is sometimes not really reasonable to apply them considering all the situations listed above that may be involved.

Check lists for the observance of regulations generally have faults and omissions relative to the fire safety measures established to protect buildings in historic areas. The regulations cannot cover every situation in these zones and even so the regulations are not always properly enforced.

Fire risk analysis methods for historic centres tend to be adaptations of the usual methods, but this does not necessarily mean that the method is suitable for the above specificities of historic centres.

Fire risk analysis may be used for construction units with more than one building. This type of concept is acceptable to historic areas, because the buildings are very close each other, without a separation wall with enough fire resistance and the adjacent roofs are with the same height. The propagation of the fire between buildings is easy.

The ranking analysis methods are used to order the situations before and after the implementation of some protective measures, comparing them with a reference value that represents the acceptable fire safety level. Constant values are attributed, to multiply each factor according its importance to the building protection or to the fire outbreak hazard and to its propagation. The parameters in these methods have to be chosen carefully, because if they are not well considered, some errors may be introduced in the calculation process. These errors are not always clear for the designer and the final result may be not correct. In case of historic areas the set of parameters to be considered is so big and complex, that this type of errors has more probability to occur [4].

Dobbernack [5] presents some fire risk ranking assessments, which indicate that the ranking methods most applicable to fire risk analysis in historic areas are:

- Risk Value Matrix Method: is the NFPA historic centre risk assessment method. This method just shows relative numbers, without any absolute result, which means that hazard and risk values within each assessment are compared. The likelihood of the event occurring and its consequences are considered. The risk takes the value of the product of fire risk value and fire hazard value. The hazard is defined for five coordinated descriptions, the fire risk for another five descriptions and these, organized into a matrix, can take 25 different values. They are compared one by one to support the decision on which combination is best, for the prevailing conditions.

- SIA 81 – the Gretener Method: is a Swiss assessment method created by Max Gretener to assess industrial buildings for insurance companies. It has been revised several times, and adapted to new situations, always focusing on the interests of the property and the insurance companies. For this reason it does not consider the need to evacuate the occupants or to protect and guarantee the activities in there. For historic centres, despite this method being widely used, buildings are generally small or partitioned and residential, so there are many factors missing from these zones that the method normally assumes. Furthermore there are many risks attached to these buildings that are not found in large industrial buildings and which the method does not consider.

- Fire Risk Assessment Method for Engineering (F.R.A.M.E) was developed from the Gretener Method to address its failings. It can be applied to planned or existing buildings, taking into account the property protection, the people's safety and the interruption of labour activities. Like the Gretener Method, the FRAME method considers some factors that are not directly applicable to historic centres and miss others that would be important, prejudicing the buildings' risk assessment. The FRAME is more favourable to the safety of the people than the Gretener method
- Hierarchical method is a Delphi method developed by Edinburgh University. An expert group is always required to select the policy, the objectives, the strategies, the parameters and the survey items and rank them according the situation, discussing all the steps and with everybody in the group agreeing the same solutions. Each member of the Delphi group must agree with all choices.

Quantifying methods like the events tree or the faults tree are more accurate and also harder to use. They list all occurrences and all their possibilities until the fire is extinguished, either by human action or because it burns itself out. The events tree goes from the beginning of the fire until the end, looking for the stages of development and the fault tree does the opposite: if the undesired event happened which step was at fault in terms of protection? It is thus possible to evaluate many ways the fire may develop, choosing the events or faults and putting them in order.

To organise the possibilities and sequences of events/faults from some initial point, the events/faults tree takes human behaviour and the reliability of the protection systems' installation into account. This kind of assessment requires specific and accurate information, only possible with research and a database on similar situations.

To build the decision trees, a general engineering project is created to choose a solution for the fire safety objectives, proving that all objectives have been met. Essentially, this requires seven steps: choosing the initial event; identifying the fire safety sources; building the tree; ranking the consequences; estimating each possibility/probability; quantifying and ranking the consequences according to the previous step, and finally evaluating the results.

The fire risk analysis in historic centres requires the definition of a lot of details, some of them difficult to obtain, because private buildings have to be examined in the public interest. The owners must be made aware of the studies and their importance if they are going to permit the visits.

Historic centres need at least two methods to give an adequate assessment in order to make the most of the potentialities of each one and nullify their faults. If it is possible to understand how and why fires do or do not occur, a more realistic analysis can be made to back a decision on what measures should be used to protect individual buildings and the entire zone.

3 The case of Montemor-o-Velho – defining scene

To better observe and understand fire risk analysis in the historic areas of the Portugal's cities, the historic zone of Montemor-o-Velho, a town in the centre of Portugal, was chosen as a case study. Montemor-o-Velho has been developed under tourist revitalization plans, and this involves better town planning. Natural elements and historic places will gain from tourism and foster local growth. Fire risk analysis can improve the plans by introducing the notion of fire safety locally, to safeguard historic buildings and their surroundings.

The area of interest is the slope of a hill on which the town's pre-medieval castle stands. The castle is the most important tourist point of the city. Of the characteristics of Portuguese historical centres listed earlier, this hill has a difficult slope, there is a difference of over 40 meters between the top and foot, representing about 4 bars, difference in the water pressure available for fire suppression, with serious implications for fire safety.

Montemor-o-Velho's historic centre has a population of about 500, most of who were born there. They are mainly elderly people. This and other social characteristics of the population of this zone must be considered, so as to propose some preventive measures against the possibility of fire in the assessments applied, especially because some measures relate to how ready the population is to act in terms of fire prevention and suppression.

A number of businesses and facilities are located in the lower zone, along the main street. These include schools and the local health centre. Further up the hill, we find more private houses, although some businesses and dwellings were paid more attention because they might represent an extra hazard in view of the factors of difficult access, lower water pressure and more chance of any fire that occurs spreading.

The zone in question has hydrants connected to the mains water system, and in the most problematic area there are not enough hydrants to cater for a fire risk, and the diameter of the water pipes is not large enough. The mains water system was remodelled a few years ago to deactivate the castle's tank. Two hydrants from the old network are still in operation, but all the others were moved to closer points in the new system. The problem here is that neither the residents nor even some firemen do not know which hydrants are working, and which are not. The hydrants are not checked or maintained regularly, and there is no guarantee that the hydrant that is needed will work well in the event of fire.

The district's fire fighting and safety is under the responsibility of the Montemor-o-Velho Volunteer Fire Brigade, and is housed outside the perimeter of the historic centre, but only about one kilometre from any part of this zone. The Fire Department has one computerised communication centre and a 24 hour alarm system. It has four large and medium water capacity vehicles, one ladder-vehicle with a 30 meter reach for rescues, two ambulances and 2 small water capacity vehicles. These may be good resources, but since only the last two can reach most of the zone, and neither can reach some parts, this does not mean a great deal. The other vehicles can only cover the main road and the main street of the historic centre, where the water supply for fire fighting is not a problem. The principal problem for the rescue vehicles, in most of the areas of the historic part, is the narrow streets that sometimes become narrower because some private vehicles park on them.

The zone has another big problem related to the storage of the domestic gas bottles. The area doesn't have piping gas systems and quite all houses and shops have gas bottles not safely stored. The shops do not have fire detection or alarm systems: they are old buildings that are not covered by the law either because are too small or because they are not inspected for law enforcement purposes.

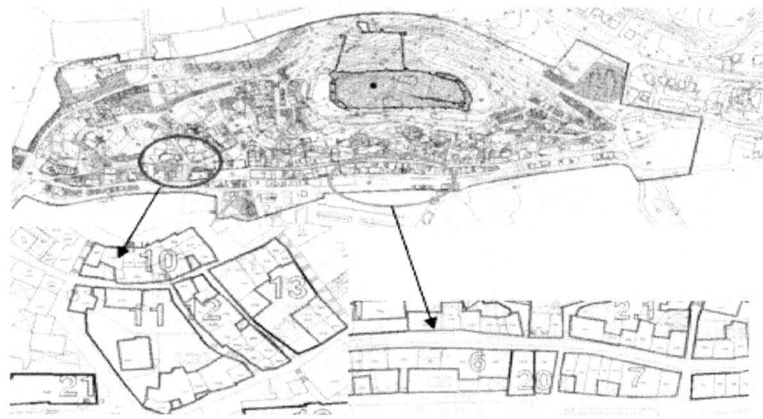

Figure 2: Montemor-o-Velho: general view with the study zones. The most critical zone marked in red and the most favourable zone marked in green.

4 The case of Montemor-o-Velho – assessments apply

The fire risk assessment methods chosen to be applied in the case of Montemor-o-Velho were the Gretener and the FRAME methods. Although these two methods are very similar with respect to their main features, this choice was intended to delimit the boundary that separates them, identifying their common and individual good and bad points, as well as trying to achieve similar solutions.

First, 21 regions were pinpointed in the historic zone, selected according to the buildings' proximity, to create uniform groups that could be dangerous perimeters in case of fire (fig. 2). Six isolated buildings were also selected, because of their size and uses. Appling the two methods to these initial assumptions revealed that the problem was more complicated. Even the more sophisticated measures for the regions could not satisfy the Gretener Method, it was worse for FRAME method. All the features of the historical centre referred above and the large areas encompassed by these perimeters made it impossible to assess some regions as safe, even if all possible measures were applied. It was therefore necessary to observe the regions more closely, one by one. They were divided into sub-regions capable of ensuring safety according to both methods, using reasonable measures. This subdivision is a first measure to be guaranteed by having structural barriers between the buildings studied. In some regions the buildings were analysed almost individually.

The fire hazards considered by Gretener and FRAME methods depend on the content in terms of goods and the construction solutions of the building. As the buildings don't have a separation between floors, with enough fire resistance, because they are made of old timber, very dry, and there isn't a staircase, the total fire load is the summation of the fire loads of all floors related to the ground-floor. When a division between floors exists, the fire load to be considered is the one of the floor with a highest value.

In the Gretener method, the combustibility, smoke and toxicity factors were defined by the most dangerous of all uses-type/goods inside the building, occupying at least 10% of the area. If there is a use-type that occupies less than 10% of the area, but has a high fire hazard in terms of the above-mentioned parameters, the value considered for, was the medium of the values suggested in the method [6].

In FRAME method the openings area considered was 5% of the perimeter.

In terms of normal protection measures, the alternatives considered to be available are, first, people training, and then the adequate portable extinguishing systems. The interior hydrants were considered unnecessary in most cases because they are very expensive and wouldn't be used to protect more than one building. This solution was only considered in specific situations. The water transport conduits needed are always shorter than the 70 meter standard length considered by the Gretener method.

Where the mains water pipes were less than 90 mm diameter, two alternatives were assessed: changing the pipes to a 120 mm diameter set, or doubling the present system by adding another 90 mm diameter set. The second option is cheaper and achieves better safety values in both methods.

The special measures initially adopted concern the fire brigade and its action delay time. In what concerns to the fire brigade, it was considered a department with all necessary equipment and at least one 1200L water capability vehicle, at least 20 trained men in the brigade who can always be contacted by phone, and with 4 men always on duty, ready to act in relation to fires and gas leak protection, and able to leave at 5 minutes notice. The fire department of Montemor-o-Velho easily verified these exigencies.

As the entire zone is less than one kilometre in diameter the delay time for the fire department to act was always held to be less than 15 minutes. Additional special measures were only considered in some special cases and only fire detection and alarm measures can be justified in the cases studied.

The sole structural measure considered reasonable in some situations was the horizontal division of floors by elements having at least 60 or 120 minutes fire resistance.

The ignition hazard was considered generally normal, except in abandoned houses where it was considered low, and one electronics shop and a disco where it was high. People's exposure to hazard was standard, except when it was possible to have more than 30 people inside the critical compartment.

Case by case, using all measures under different combinations, it was possible to reach the property safety according both methods applied. In what concerns to the people's safety, the Gretener solutions, when using the FRAME method, were not enough to obtain the desired safety level.

The extra measures to reach the necessary safety level were applied to the zone as a whole, considering a big building and all existent buildings as compartments. Some of the measures were applied to the buildings itself (such as energy sources like lighting, heating and gas) and the others to the zone (such as hydrants, emergency signs, safety routes and safety plans).

5 Conclusions

The Gretener method is sensitive to changes in water supply and compartment conditions. Accurate information about water pressure, pipe diameter and water tank size can make difference to using this method. Another important factor in this method is whether the integrity of the compartmention can be guaranteed by means of 30 or 60 minute fire resistance elements. The Gretener method is not sensitive to the changing of the area, for certain relations between the length and the width of the fire compartment.

The FRAME method is sensitive to area changes that benefit small buildings.

In addition FRAME method can evaluate sites' continuity of activities, a point not considered here.

The properties of interior goods, in the FRAME method, are considered quantitatively, using a critical temperature, combustible dimension and combustibility rate with numerical values, contrary to Gretener method which classifies not only combustibility but also smoke and toxicity using qualitative concepts as low, medium or high, and then assumes factor values for them.

The changes in compartmention conditions with respect to structural fire resistance are not visible in the FRAME method.

To achieve people's safety in the FRAME method, at least half of all normal and safety route measures must be adopted.

Both the methods applied show a lot of deficiencies and faults for the analysis of historic centres, but they also show how important it is that some analysis is undertaken. They expose the faults that can be seen easily but need some support if they are to be remedied and also less obvious faults that are equally dangerous. The Gretener and FRAME methods complement one another for isolated buildings and small regions; for a zone the application of both is still not enough to analyse all questions involved in historic centre assessment, but if extra measures are taken, known from expertise to establish the situation as a "small region", the methods meet all expectations.

References

[1] Câmara Municipal de Guimarães – *Plano Piloto de Luta contra Incêndios e Segurança.* Gabinete Técnico Local. 2005.

[2] Coelho, A. L. – *Segurança contra Risco de Incêndio em Áreas Urbanas Antigas – Princípios Gerais de Intervenção.* Seminário "Riscos e Vulnerabilidades em Centros Urbanos Antigos". Évora, Portugal, 2000.

[3] Gonçalves, J. M. F. – *Incêndios em Núcleos Urbanos Antigos Verificação da Segurança contra Incêndios na Mouraria.* MSc Thesis, Universidade Técnica de Lisboa, Instituto Superior Técnico, 1994.

[4] De Smet, E. - *Handbook for the use of this Fire Risk Assessment Method for Engineering,* 2nd. Edition, 1999.

[5] Dobbernack, R. – *Fire Risk Assessment Methods.* Final Report of Workgroup 6. Fire Risk Evaluation to European Cultural Heritage – FIRE-TECH, 2003.

[6] Neves, I. C; Tovar De Lemos, A. M. F. – *Avaliação do Risco de Incêndio – Método de Cálculo.* Translation and adaptation of the original SIA publication "Brandriskobewertung Berechnungsverfahren", 1987.

Lateral-torsional buckling of ferritic stainless steel beams in case of fire

P. Vila Real[1], N. Lopes[1], L. Silva[2] & J.-M. Franssen[3]
[1]*Department of Civil Engineering, University of Aveiro, Aveiro, Portugal*
[2]*University of Coimbra, Coimbra, Portugal*
[3]*Department ArGEnCo, University of Liege, Liege, Belgium*

Abstract

This work presents a numerical study of the behaviour of ferritic stainless steel I-beams subjected to lateral-torsional buckling and compares the obtained results with the beam design curves of Eurocode 3. New formulae, for the lateral-torsional buckling, that approximate better the real behaviour of ferritic stainless steel structural elements in case of fire are proposed. These new formulae were based on numerical simulations using the program SAFIR, which was modified to take into account the material properties of the stainless steel.
Keywords: ferritic, stainless steel, Eurocode 3, numerical modelling, lateral-torsional buckling, fire.

1 Introduction

There are five basic groups of stainless steels, classified according to their metallurgical structure: the austenitic, ferritic, martensitic, duplex austenitic-ferritic and precipitation-hardening groups [1]. Austenitic stainless steels provide a good combination of corrosion resistance, forming and fabrication properties. Duplex stainless steels have high strength and wear resistance with very good resistance to stress corrosion cracking. The most commonly used grades, typically referred to as the standard austenitic grades, are 1.4301 (widely known as 304) and 1.4401 (widely known as 316). The austenitic stainless steels are generally the more useful groups for structural applications but increasing interest in ferritic steels for structural purposes has been recently noted due to its low cost. The responsibility of the final cost of the austenitic stainless steel is the price of nickel. Typically they contain 8.0-13.0% of nickel whereas ferritic

WIT Transactions on Engineering Sciences, Vol 58, © 2007 WIT Press
www.witpress.com, ISSN 1743-3533 (on-line)
doi:10.2495/EN070101

stainless steels contain a low nickel level. The ferritic stainless steel 1.4003 studied in this work contains 0.3-1.0%.

The biggest advantage of stainless steel is its higher corrosion resistance. However, its easy maintenance, high durability and reduced life cycle costs are also important properties. It is also known that the fire resistance of stainless steel is higher than the carbon steel usually used in construction.

EN 1993-1-4 "Supplementary rules for stainless steels" [2] gives design rules for stainless steel structural elements at room temperature, only making mention to its fire resistance by doing referring to the fire part of the Eurocode 3, EN 1993-1-2 [3].

Although its use in construction is increasing, it is still necessary to develop the knowledge of its structural behaviour. Stainless steels are known by their non-linear stress-strain relationships with a low proportional stress and an extensive hardening phase. There is not a well defined yield strength, being usually considered for design at room temperature the 0.2% proof strength, $f_y=f_{0.2proof}$. In a fire situation higher strains than at room temperature are acceptable, so Part 1.2 of Eurocode 3 suggests the use of the stress at 2% total strain as the yield stress at elevated temperature θ, $f_{y,\theta}=f_{2,\theta}$, for Class 1, 2 and 3 cross-sections and $f_{y,\theta}=f_{0.2proof,\theta}$, for Class 4. Comparison of the reduction of strength and elastic stiffness of structural carbon steel and stainless steel at elevated temperature for several grades of stainless steels (as defined in EN 1993-1-2 [3]) is shown in figures 1 and 2, where $k_{y,\theta}=f_{y,\theta}/f_y$ and $k_{E,\theta}=E_\theta/E$, being $f_{y,\theta}$ and f_y the yield strength at elevated temperature and at room temperature respectively, and E_θ and E the modulus of elasticity at elevated temperature and at room temperature.

The stainless steel mechanical and thermal properties at high temperatures, used in this paper, can be found in Part 1-2 of Eurocode 3 [3]. For the evaluation of the yield strength reduction factor, the Eurocode states that the following equation should be used:

$$k_{y,\theta}=\left[f_{0.2p,\theta}+k_{2\%,\theta}\left(f_{u,\theta}-f_{0.2p,\theta}\right)\right]\frac{1}{f_y} \tag{1}$$

where

$f_{0.2p,\theta}$ is the proof strength at 0.2% plastic strain, at temperature θ;
$k_{2\%,\theta}$ is the correction factor for determination of the yield strength $f_{y,\theta}$;
$f_{u,\theta}$ is the ultimate tensile strength, at temperature θ.

Despite both carbon and stainless steel exhibiting different constitutive laws, whereby stainless steel presents a pronounced non-linear behavior even for low stress values, the stainless steel design rules are based on those developed for carbon steel. In a previous paper [4] a new proposal for the lateral-torsional buckling of austenic grades stainless steel beams was made. In the present paper a similar study is done for the ferritic stainless steel grade 1.4003 (the only ferritic stainless steel presented in Part 1.2 of the Eurocode 3).

Figure 1 shows that the variation of the strength reduction of the stainless steel grade 1.4003 with temperature is different from the other grades, mainly for the temperature range from 500°C to 700°C.

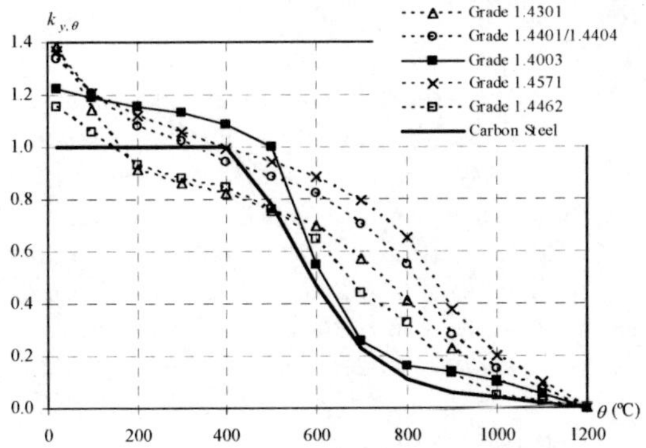

Figure 1: Strength reduction at high temperatures.

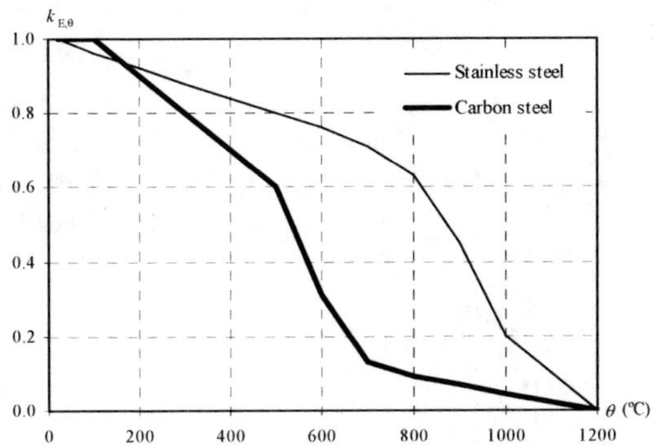

Figure 2: Elastic stiffness reduction at high temperatures.

The reduction of the yield strength and the reduction of the modulus of elasticity are used in the determination of the non-dimensional slenderness at high temperatures, as it will be shown later in this work.

Program SAFIR [5], a geometrical and material non linear finite element code, which has been adapted according to the material properties defined in EN 1993-1-4 [2] and EN 1993-1-2 [3], to model the behaviour of stainless steel structures [6] has been used in the numerical simulations. This program, widely used by several investigators, has been validated against analytical solutions, experimental tests and numerical results from other programs, and has been used in several studies that lead to proposals for safety evaluation of structural elements, already adopted in Eurocode 3. In the numerical simulations, geometrical imperfections and residual stresses were considered [4].

The objective of the study presented in this paper is to evaluate the accuracy of the lateral-torsional buckling design procedures prescribed in Eurocode 3, for I cross-sections in stainless steel grade 1.4003, at high temperatures. This study concluded that the Eurocode 3 formulae need to be improved and that a new proposal should be made for the stainless steel grade 1.4003.

2 Case study

A simply supported beam with fork supports subjected to uniform bending diagram was chosen to explore the validity of the beam safety verifications.

The influence of the cross-sectional shape, assessed using the height/width (h/b) relation, was taken into account in this work. Equivalent welded cross-sections equivalent to an IPE220 (representative of h/b = 2), HE500A (representative of h/b < 2) and IPE500 (representative of h/b > 2) were studied.

A uniform temperature distribution in the cross-section was used so that comparison between the numerical results and the Eurocode could be made. In this paper, the temperatures chosen were 400, 500, 600 and 700 ºC, deemed to cover the majority of practical situations.

In the numerical simulations, a lateral geometric imperfection given by the following expression was considered:

$$y(x) = \frac{l}{1000}\sin\left(\frac{\pi x}{l}\right) \tag{2}$$

where l is the length of the beam. An initial rotation around the beam axis with a maximum value of $l/1000$ radians at mid span was also considered.

The adopted residual stresses follow the typical patterns for carbon steel welded sections, considered constant across the thickness of the web and flanges. The distribution is shown in Figure 3, and has the maximum value of f_y (yield strength) [7–9].

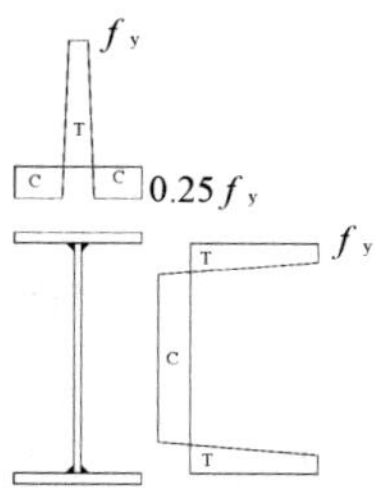

Figure 3: Residual stresses: C – compression; T – tension.

3 Formulae for lateral-torsional buckling

3.1 Eurocode 3 formulae for stainless steel elements

For stainless steel beams subjected to high temperatures, Part 1-4 of Eurocode 3 [2] states that the same formulation prescribed for carbon steel elements should be used, according to the EN 1993-1-2 [3], where the lateral-torsional buckling resistance for class 1 and class 2 cross-sections is given by

$$M_{b,fi,t,Rd} = \chi_{LT,fi} W_{pl,y} k_{y,\theta} f_y \frac{1}{\gamma_{M,fi}} \tag{3}$$

where

$$\chi_{LT,fi} = \frac{1}{\varphi_{LT,\theta} + \sqrt{(\varphi_{LT,\theta})^2 - (\bar{\lambda}_{LT,\theta})^2}} \tag{4}$$

with

$$\varphi_{LT,\theta} = \frac{1}{2}\left[1 + \alpha\bar{\lambda}_{LT,\theta} + (\bar{\lambda}_{LT,\theta})^2\right] \tag{5}$$

In this expression the imperfection factor α depends on the steel grade and is given by

$$\alpha = \beta\sqrt{235/f_y} = 0.65\sqrt{235/f_y} \tag{6}$$

The non-dimensional slenderness for lateral-torsional buckling at high temperatures is given by

$$\bar{\lambda}_{LT,\theta} = \bar{\lambda}_{LT}\left[\frac{k_{y,\theta}}{k_{E,\theta}}\right]^{0.5} \tag{7}$$

being

$$\bar{\lambda}_{LT} = \sqrt{\frac{W_{pl,y} f_y}{M_{cr}}} \tag{8}$$

where $W_{pl,y}$ is the plastic bending modulus, f_y is the yield strength of steel and M_{cr} is the elastic critical moment for lateral-torsional buckling.

3.2 Proposal for austenitic and duplex stainless steel elements

Figure 4 shows that the stainless steel beam design curve for lateral-torsional buckling from Eurocode 3 is not on the safe side. To improve this curve new severity factors β, given in table 1, different from the one used for carbon steel (see equation 6), were proposed by the authors [4, 6]. The new severity factor takes into account the influence of the shape of the cross-section. In this previous work the authors did not consider the influence of the stainless steel grade.

4 Comparison between the lateral-torsional buckling formulae and the numerical results

Application of eqs. (3) to (8) and Table 1 to ferritic stainless steels leads to the results of Figure 5, that compare the numerical results and the code proposals. It is clear that this proposal, based on austenitic stainless steels, is not accurate for ferritic stainless steels.

In this figure it can be observed that a beam with a length of 5 m exhibits slenderness values for 700ºC and 600ºC quite different from the corresponding values for 400ºC and 500ºC. These differences are not so big for the case of austenitic stainless steel as shown in fig. 4. These differences result from the reduction of the yield strength (see fig. 1) is shown in fig. 6. As it can be seen in equation (7) the slenderness at room temperature is multiplied by the factor $(k_{y,\theta}/k_{E,\theta})^{1/2}$ in order to obtain the slenderness at high temperatures. Figure 6 shows that from 500°C to 700°C there is a great decrease of this factor for the 1.4003 stainless steel, which does not occur with the others stainless steel grades.

Table 1: Values of the severity factor β.

Cross-section	Limits	β
Welded I section	h/b ≤ 2	0.85
	h/b > 2	1.00

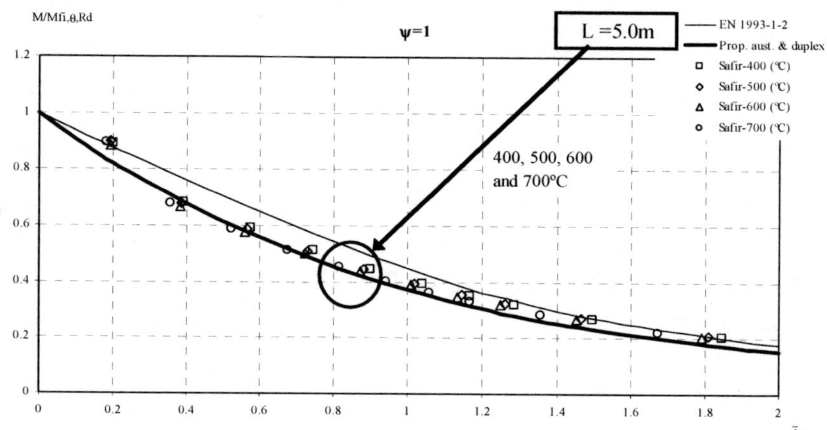

Figure 4: Lateral-torsional buckling in IPE 500 beams of the stainless steel grade 1.4301.

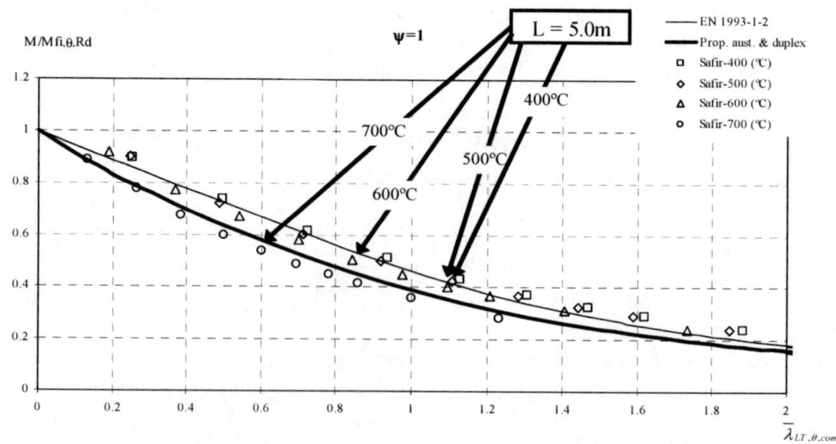

Figure 5: Lateral-torsional buckling for in IPE500 beams of the stainless steel grade 1.4003.

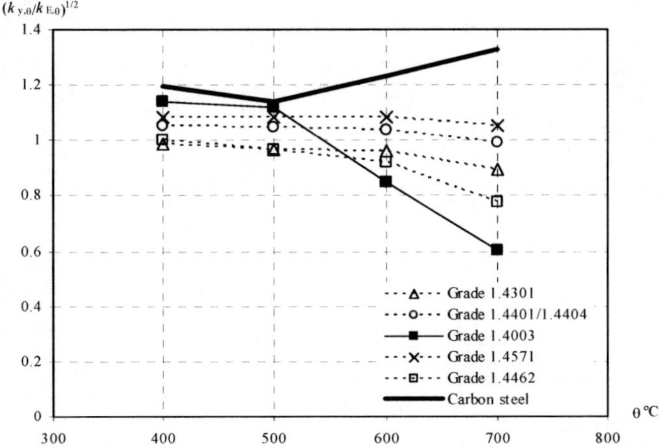

Figure 6: Variation of the square root used in the determination of the slenderness.

From figure 5 it can be concluded that the previous proposal [4] and the Eurocode 3 are not safe for the case of the ferritic stainless steel grade 1.4003. In the next section a new proposal covering ferritic stainless steel grades will be presented.

5 New proposal for the ferritic stainless steel grade 1.4003

Based on a parametric study considering the influence of the shape of the cross-section a new severity factor β, given in table 2, was found for the stainless steel grade 1.403.

Table 2: Values of the severity factor β for the 1.4003.

Cross-section	Limits	β
Welded I section	$h/b \leq 2$	1.00
	$h/b > 2$	1.20

Figures 7 to 9, compare the beam design curves obtained using Part 1-2 of Eurocode 3, described in section 3.1 of this paper (denoted "EN 1993-1-2"), the curve obtained with the new severity factor given in table 2 (denoted "New proposal"), and the numerical results obtained with the program SAFIR.

Table 3 summarizes the new severity factor taking into account the influence of the type of the stainless steel.

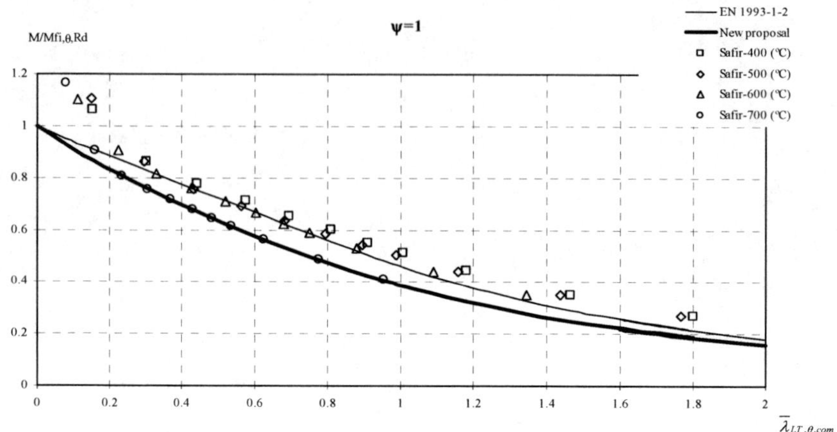

Figure 7: Lateral-torsional buckling in HE500A beams (representative of h/b < 2) of the stainless steel grade 1.4003.

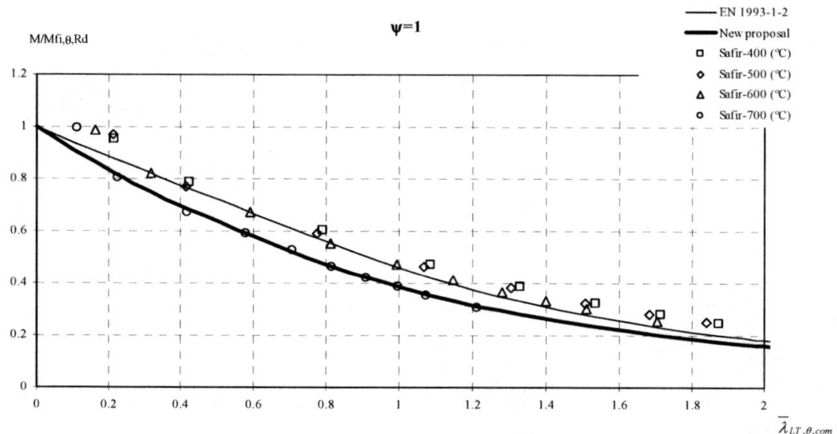

Figure 8: Lateral-torsional buckling in IPE220 beams (representative of h/b = 2) of the stainless steel grade 1.4003.

6 Conclusions

This paper has shown that the previous proposal made by the authors [4], for the lateral-torsional buckling resistance of unrestrained stainless steel beams under fire loading, based on the austenitic stainless steel, is not safe for the ferritic stainless steel grade 1.4003. A new severity factor that takes into account the influence of the steel grade, as well as the influence of the slenderness of the cross-section (relation h/b of the cross-section) has been proposed being in good agreement with the numerical results obtained with the program SAFIR. This study also has shown that the slenderness of the cross-section should be taken into account as it is already proposed in Eurocode 3 for carbon steel elements at room temperature.

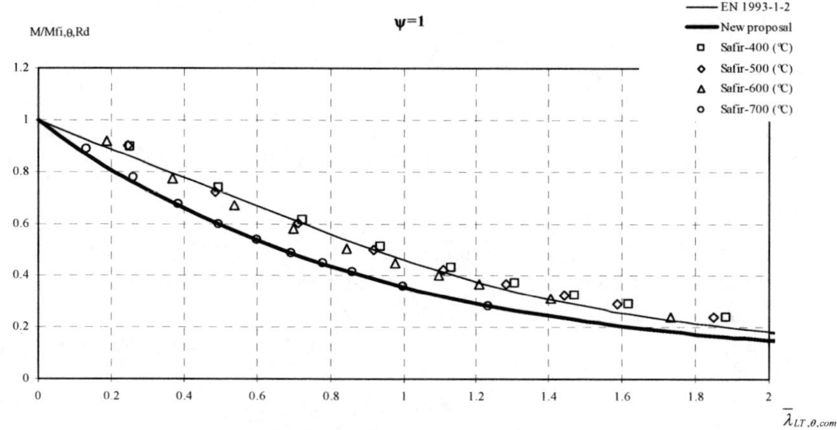

Figure 9: Lateral-torsional buckling in IPE500 beams (representative of h/b > 2) of the stainless steel grade 1.4003.

Table 3: New proposal for the severity factor β.

Cross-section	Limits	β	
		Austenitic and Duplex stainless steel	Ferritic stainless steel 1.4003
Welded I section	h/b ≤ 2	0.85	1.00
	h/b > 2	1.00	1.20

Acknowledgement

The authors wish to acknowledge the Calouste Gulbenkian Foundation (Portugal) for its supports through the scholarship given to the second author.

References

[1] Euro Inox e Steel Construction Institute "Design Manual for Structural Stainless Steel", 3rd edition, 2006.

[2] European Committee for Standardisation, prEN 1993-1-4 "Eurocode 3, Design of Steel Structures – Part 1-4. General rules – Supplementary Rules for Stainless Steels", Brussels, Belgium, 2005.

[3] European Committee for Standardisation, EN 1993-1-2 "Eurocode 3: Design of Steel Structures - Part 1-2: General rules - Structural fire design", Brussels, Belgium, April 2005.

[4] Lopes, N.; Vila Real, P.; Silva, L.; Franssen, J.-M.; Mirambell, E. "Proposal to the Eurocode 3 for the lateral-torsional buckling of Stainless steel I-beams in case of fire" actas do Fourth International Workshop Structures in Fire SiF'06, pp. 463-472, ISBN 84-95999-74-9, Aveiro, Portugal, 10 to 12 of May of 2006.

[5] Franssen, J.-M. 2005. SAFIR. A Thermal/Structural Program Modelling Structures under Fire. Engineering Journal, A.I.S.C., Vol. 42, No. 3, pp. 143-158.

[6] Lopes, N., Vila Real, P.M.M., Piloto, P., Mesquita, L. e Simões da Silva, L., "Modelação numérica da encurvadura lateral de vigas I em aço inoxidável sujeitas a temperaturas elevadas", (in portuguese) Congreso de Métodos Numéricos en Ingeniería, Granada, Spain, 2005.

[7] Chen W. F. and Lui E. M. Stability design of steel frames. CRC Press, 1991.

[8] Gardner, L., Nethercot, D. A. 2004. Numerical Modeling of Stainless Steel Structural Components - A consistent Approach. Journal of Constructional Engineering, ASCE, pp. 1586-1601.

[9] Greiner, R.; Hörmaier, I.; Ofner, R.; Kettler, M., 2005. Buckling behaviour of stainless steel members under bending. ECCS Technical Committee 8 – Stability.

Rise in brittle fracture resistance of spherical vessels for storage of liquid ammonia under extreme forces

E. M. Basko, V. V. Larionov & V. N. Lazutin
Melnikov institute of steel structures, Moscow, Russia

Abstract

This report presents the results of spherical vessel investigation and laboratory research of physico-mechanical properties of steel and welded joints. Dynamics of corrosion-mechanical cracking of the vessel welded joints and brittle fracture resistance at low temperatures and dynamic forces have been given in the context of fracture mechanics. The major causes of corrosion-mechanical cracking of the vessel welded joints in the medium of liquid ammonia have been determined. The performance of works on stress deconcentration in the butt-welded joints on the inside of the wall being in contact with ammonia was demonstrated to be an effective method of increasing resistance to corrosion-mechanical cracking and brittle fracture.
Keywords: spherical vessel, ammonia, pressure, cracks, defects, welded joints, life, cracking resistance, brittle fracture, dynamic stress.

1 Introduction

Practice of application of the welded pressure vessels for storage of liquid ammonia demonstrated that corrosion-mechanical cracking (CMC) of welded joints is made possible in the course of their service. The action of extreme forces in cracking can result in brittle fracture of vessels at stresses that are substantially below the design values. This is associated both with the resulting dynamic effects and with dependence of cracking resistance characteristics and critical temperatures of vessel material brittleness on strain rate. In this connection when evaluating resistance to fracture of vessels for storage of ammonia on action of extreme forces it is necessary to take into account

WIT Transactions on Engineering Sciences, Vol 58, © 2007 WIT Press
www.witpress.com, ISSN 1743-3533 (on-line)
doi:10.2495/EN070111

damages accumulating during service, on the one hand, and minimize formation and growth of cracks during the design period of service by use of the suitable materials and technologies, on the other hand.

Russia first faced with a CMC problem for the spherical vessels with a volume of 2000 m^3, in which storage of liquid ammonia is carried out at the excessive pressure of 0.6 MPa and a temperature as high as 12 °C, in the middle 80s when during their inspection cracking of the welded joints was discovered and several vessels failed ahead of schedule (Pohodnya *et al.* [1], Basko *et al.* [2]).

In this connection the institute accepted a program, on CMC-problem research which included execution of full-scale investigation of technical condition of liquid ammonia pressure vessels, study of physico-mechanical properties of vessel material and development of methods for increase of resistance to corrosion-mechanical cracking and brittle fracture taking into account the effect of low temperatures and dynamic effects.

The effect of low temperatures and dynamic loading are considered as the extreme cases exceeding the bounds of design conditions of vessel service. Decrease in temperature of wall to minus 30-40 °C is made possible in case of break of wall thermal insulation and depressurization of wall and leakage of ammonia can result in its local supercooling up to minus 67 °C.

The dynamic loads are possible as a result of the external actions of seismic or wind character, et al. as well as initiation of a brittle crack in the zone of local supercooling of the wall section during its depressurization, action of which is comparable with the external dynamic loading by its effect (Basko *et al.* [3]).

2 Analytical treatment of investigation results

Since 1990 there was carried out an investigation of technical condition of 12 spherical vessels after a lapse of 6, 8, 12 and 24 years of their operation. There was performed inspection of the base metal and welded joints on the inside of the wall being in contact with liquid and gaseous ammonia.

The visual inspection of the welded joints showed a great deal of technological defects in the form of gas pockets, chains of undercuts, flux inclusions, high ripple of weld beads, et al. practically in all vessels. The primary inspection of the welded joints by methods of color and ultrasonic test did not discover cracks.

To increase a sensitivity of capillary method on one of the first vessels, there was carried out the mechanical removal of the weld bead bulge with the subsequent grinding flush with the base metal. First of all such work was performed in the butt-welded joints with technological welding defects. Inspection of the welded joints prepared by the mentioned above way demonstrated a rich variety of the surface cracks. The crack width was equal to 6-10 microns and this fact required the careful preparation of the welded joints surfaces for their detection. In this connection later on inspection of the welded joints was carried out only after the mentioned mechanical preparation of welds, length of which in one spherical vessel amounts to $\approx$700 m.

Inspection of the welded joints with the following grinding of welds showed the surface cracks in all investigated vessels. All cracks were activated because of the welding defects. Length and number of cracks depended on type and size of the welding defects as well the vessel service life. The lowest number of cracks (20 pcs 3-15 mm in length) was discovered in the spherical vessel after 6 years of its service [2]. At the same time the welded joints of this vessel had the lowest number of technological defects allowable for welding. The welding cracks were oriented both along and across axis of welds. The length of cross cracks was no more than 50 mm and that of longitudinal cracks was as high as 2 m [2]. The depth of longitudinal cracks was no more than 8 mm whereas that of cross cracks reached 18 mm. Cases of the wall depressurization associated with crack propagation were not registered. The typical distribution of cracks throughout the depth obtained on the basis of inspection results of 2 spherical vessels after 24 years of their service is presented in fig.1; in each of these vessels more than 270 cracks were discovered.

As evident from the cited data, more than 50% of the surface cracks were as deep as 2 mm. These cracks were formed from small defects of welds up to 0.5 mm in depth. Only 5% of cracks had a depth of more than 2 mm. The maximum depth of cracks reached a value of 17-18 mm. At the same time the similar presence of defects is made also possible for a smaller service life, but with a larger quantity and sizes of technological defects. The analogous data on damageability of 4 spherical vessels after 16 years of their service are presented in the work of Kuzeev [4]; according to its information more than 160 surface cracks were discovered in each of these vessels.

The corrosion-mechanical cracking of welds in the liquid ammonia medium occurs in the cylindrical pressure vessels too (Basko *et al.* [2]). However in connection with a far less length of welds, on the one hand, and a lower level of stresses resulting from internal pressure and higher quality of welds, on the other hand, the total quantity of these cracks is significantly less than in the spherical vessels.

Following the primary inspection of the welded joints all discovered defects were eliminated by a mechanical way. The defective parts of the welded joints with a depth of crack sampling more than 3 mm were subjected to backing run, grinding flush with the base metal and inspection. After the hydraulic test operation all the investigated vessels were approved for their service.

In future there was monitoring of four investigated spherical vessels. In accordance with it the careful inspection of the welded joints for defect detection was carried out every 2–3 years of their service. As a result of repeatedly executed inspection it was determined that following the technological operation of stress deconcentration in the butt-welded joints the corrosion-mechanical cracking was practically arrested. The discovered individual cracks to 10–20 mm in length resulted from the small cracks missing during the primary inspection or those that were incompletely removed during the initial repair. The last investigation of spherical vessel that was carried out following 29 years of its service and 15 years after its primary inspection and repair confirmed efficiency of the executed technological operations providing considerable increase in resistance to corrosion-mechanical cracking of welds in the ammonia medium. Following investigation of mechanical properties of material cut from the wall of the given vessel and calculations of the service life and resistance to brittle fracture it was allowed to the further service for a period up to 2014 years.

3 Research results of physico-mechanical properties of the spherical vessel material

To evaluate effect of the long-term exposure of ammonia on the physico-mechanical properties of 09G2S steel (spr Russian standard) and welded joints, there was carried out cut-out of the wall fragments from the spherical vessels after 8, 12 and 29 years of their service. The results of vessel material investigation were compared with the corresponding data for 09G2S steel sheets and welded joints.

The results of laboratory research of the chemical analysis, mechanical properties and microstructure of the vessel material showed that all characteristics satisfy the requirements of standards for base metal and welded joints (Basko *et al.* [2]). The similar results were obtained in the course of the vessel material investigation after 29 years of its service.

As follows from the results of experimental investigations, the values of critical stress intensity factors K_c obtained by the results of vessel material tests are no less than the corresponding characteristics gained in testing samples made from sheets and butt-welded joints non-subjected to ammonia action. The successful welding of inserts into the walls of spherical vessels instead of cutout fragments confirmed preservation of the good metal weldability. This fact is an indirect evidence of steel hydrogenation absence in process of continuous service, as also agreed with the research results of hydrogen analysis of metal presented in the work of Basko *et al.* [2].

The performed laboratory research testifies the possible usage of mechanical properties and cracking resistance characteristics obtained in testing the spherical vessel material non-subjected to action of liquid ammonia for calculation of the load-carrying capacity of brittle fracture resistance of vessels aimed for storage of liquid ammonia.

4 Design estimation of the spherical vessel service life

As follows from the investigation results, formation of cracks in the welded joints of spherical vessels is possible already at the early stage of their operation. In process of propagation they can reach the values approaching thickness of the vessel wall. At the same time at the design stage the strength calculations are carried out without taking presence of cracks into consideration. This introduces uncertainty as into the actual load-carrying capacity of wall as in fixing the safe operation life of vessels.

On the basis of analysis of damageability dynamics for the welded joints of spherical vessels for 6–30 years of their service and performed investigations of mechanical properties and cracking resistance characteristics of vessel material

we carried out a design estimation of resistance to corrosion cracking and brittle fracture with use of methods and criteria of fracture mechanics.

In accordance with points of the fracture mechanics time T to formation of crack from the source defect a_o to the calculated value a_p can be represented in the form of eqn (1):

$$T = T_z + T_p \tag{1}$$

where T_z – time to crack birth from a technological defect;
T_p – time of crack growth from a_o to a_p.

Since, as the results of investigation indicated, time to the crack birth can be ranged from 0 to 5 years with existing technology of welding and quality of the welded joints, calculations of the service life assume to take $T_z = 0$. Then the design service life is determined by time of crack growth from a_o to the accepted design size a_p and according to Karzov *et al.* [5] it can be calculated by integration of eqn (2).

$$T_p = C_\kappa \int_{a_o}^{a_p} \Delta K^{-y}\, da \tag{2}$$

where ΔK – difference of stress intensity factors;
a_o and a_p – initial and design sizes of crack;
C_κ – empirically determined coefficient characterizing resistance to corrosion-mechanical cracking in the liquid ammonia medium.

Determination of C_κ in laboratory conditions is a very intricate and labor-intensive problem connected with assurance of test conditions approaching the actual tests over a long period of time. On the basis of damage dynamics analysis of spherical vessels made from 09G2S low-alloy steel and exploited at the operating pressure up to 0.6 MPa we obtained $C_\kappa = (1.8\text{–}2.0)\cdot 10^2$ MPa $\sqrt{m}$. The empirically determined relationship for T_p can be represented in the form of eqn (3):

$$T_p = \frac{C_\kappa}{\sigma}\left[\frac{1}{M_o\sqrt{a_o}} - \frac{1}{M_p\sqrt{a_{op}}}\right] \tag{3}$$

where σ – design stresses in the wall of vessel from the operating pressure;
M_o and M_p – coefficients taking the relative sizes and form of crack into account.

Fig.1 presents the results of calculations according to eqn (3) with different sizes a_o of source defects for the spherical vessels with a volume of 2000 m^3 at the operating stress σ = 120 MPa. The measured values of the crack depth in the vessel welded joints are also presented here (marked by daggers). As follows from Fig.1, the design curves reflect the total kinetics mechanisms of the welded joint damageability for the spherical vessels. The discovered damageability to

(0.5–0.9)t at the early stage of service of 8–12 years is concerned with presence of the source technological defects 1.5–2.5 mm in depth (curves 4–6). In the presence of the source defects 0.5–1.0 mm deep the similar damageability takes place after 20–25 years of service (curves 1–3). With defects less than 0.5 mm deep the depth of cracks for service up to 24 years does not exceed 0.2t at t = 20 mm.

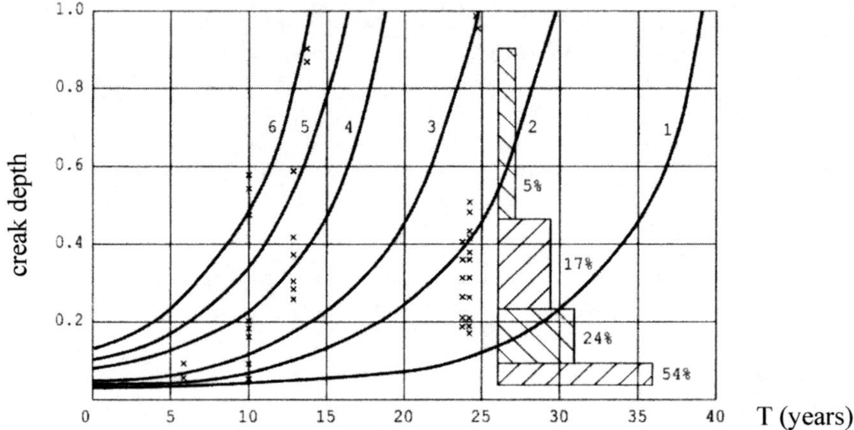

Figure 1: Diagram “Crack depth-time of service T (years)” (- distribution of crack depth after 24 years of service).

To analyze resistance to brittle fracture and service life of spherical vessels taking size and type of defects, stress level, temperature and dynamic forces into consideration, it is more convenient to represent eqn (3) in the form of eqn (4):

$$T_p = C_\kappa \frac{\Delta K}{K_o - K_p} \tag{4}$$

where ΔK – difference of stress intensity factors K_o and K_p;
K_o and K_p – initial and design values of stress intensity factors.

Formation of through-the-thickness crack results in depressurization of wall and so this fact eliminates possibility of the further service of vessel. Therefore accepting a safety factor as a value of K_p corresponding to formation of through-the-thickness crack and replacing K_p by $\sigma\sqrt{\pi t}$ from eqn (4) we obtain an expression for a design service period of spherical vessel in the form of eqn (5):

$$T_p = C_\kappa \left(\frac{1}{K_o} - \frac{n_\kappa}{\sigma\sqrt{\pi t}} \right) \tag{5}$$

where $n_k = 2$ – safety factor;
σ – operating stress in the wall of vessel;
t – thickness of wall.

Substituting the numerical values K_o = 4.0 MPa (at a_o = 0.5 mm), σ = 120 MPa, t = 20 mm, n_κ = 2 and C_κ = 180 into the eqn (5) we obtain that the design service life is equal to 33 years according to criterion of brittle fracture resistance at $a_o \leq$ 0.5 mm.

As it follows from the eqn (6), increase in CMC resistance may be reached by decrease in size a_o of the source defects, reduction of operating stresses σ by means of the operating pressure reduction or increase in thickness of the vessel wall.

5 Evaluation of brittle fracture resistance

As mentioned above, decrease in temperature of the vessel wall is possible as a result of break of the vessel thermal insulation or depressurization of wall. In the last case temperature of wall on the local section of depressurization zone can fall to minus 67 °C. The results of tension test of samples with through-the-thickness cracks showed that under the static loading transition of samples to brittle fracture at stresses that are less than yield strength takes place at temperatures below minus 80 °C. At a temperature of minus 70 °C the critical stress intensity factor K_c is no less than 85MPa$\sqrt{m}$ (Basko *et al.* [2]). With respect to the value of K_p = 30 MPa$\sqrt{m}$ corresponding to formation of through-the-thickness crack the safety factor n_k as to the critical value of K_c is no less than 2.8 and with respect to the allowable value of $[K_p]$ it is more than 5.0. This is evidence of the adequate supplies of load-carrying capacity of vessels as to criterion of brittle fracture resistance at the static loading.

The action of dynamic loads caused by the external and internal factors noted in introduction results in increase in the strain rate of material in the crack tip. In this case the critical temperatures of transition from tough to quasibrittle and brittle fractures can essentially rise. The feature of low-alloy steel behavior under conditions of dynamic load action lies in sensitivity of their mechanical properties and cracking resistance characteristics to the strain rate. At the same time the influence of strain rate becomes apparent in different ways depending on capability for plastic deformation. In case of preservation of capability for plastic deformation the yield strength of low-alloy steel increases. The value of critical stress intensity factor $K_{C\dot{\varepsilon}}$ is calculated by eqn (6) (Basko [6]):

$$K_{C\dot{\varepsilon}} = K_{C\dot{\varepsilon}_0} \frac{\sigma_{T\dot{\varepsilon}}}{\sigma_{T\dot{\varepsilon}_0}} \tag{6}$$

where $K_{C\dot{\varepsilon}}$ and $K_{C\dot{\varepsilon}_0}$ - values of K_C at the strain rates $\dot{\varepsilon}$ and $\dot{\varepsilon}_0$.

During dynamic loading the strain rate can reach the value of 10^2–10^3 s^{-1}. The yield strength of low-alloy steels at the given strain rates increases by 1.5–1.7 times. The cracking resistance limit of steel under dynamic loading increases too Basko and Makhutov [7]. Therefore in the field of tough and quasitough conditions under action of dynamic loads the brittle fracture resistance is provided.

As follows from the results of impact resistance test of samples and double tension test of large-size flat samples with initiation of brittle cracks (Basko *et al.* [3]), the critical temperature of transition from quasitough to quasibrittle fractures for 09G2S steel under dynamic loading is equal to minus 10-20 °C. Therefore at the temperatures of vessel wall below minus 10 °C the evaluation of resistance to brittle fracture taking dynamic forces into account is to be carried out with use of dynamic characteristics of cracking resistance.

According to Basko and Makhutov [7] characteristics of dynamic cracking resistance of 09G2S steel are K_d = 20 MPa$\sqrt{m}$ and 30 MPa$\sqrt{m}$ at temperatures of minus 70 °C and minus 40 °C respectively. At the given values of dynamic cracking resistance limit K_d the brittle fracture resistance of vessels under dynamic forces is provided prior to their depressurization at the wall temperatures above minus 40 °C. In case of wall depressurization and decrease of temperature to minus 67 °C on the local section of zone of through-the-thickness crack formation the crack arrest is possible at the wall temperature outside the supercooled section above minus 10 °C on condition that length of brittle crack is expressed by eqn (7) (Basko and Makhutov [7]):

$$2l_d \leq \frac{2}{\pi}\left(\frac{K_d}{\sigma \cdot n_\kappa}\right) \tag{7}$$

where $2l_d$ – length of brittle crack;
K_d – dynamic limit of cracking resistance;
σ – stress in the wall of vessel;
n = 1,2 – safety factor as to the dynamic cracking resistance limit.

Substituting the corresponding values into eqn (7) we obtain the crack length $2l_d$ < 110 mm at which the brittle fracture resistance of spherical vessel is provided in case of local depressurization of wall.

6 Conclusion

The results of performed investigations are evidence of practical expediency and applicability of the fracture mechanics concepts both for dynamics analysis of corrosion-mechanical cracking, evaluation of brittle fracture resistance of the steel vessels taking a stressed state, quality of manufacturing and service conditions into account and analysis of extreme cases associated with action of low temperatures and dynamic loading. The obtained results of investigations made it possible to formulate the basic positions on calculation of service life and evaluation of brittle fracture resistance of the spherical vessels as well as to propose an effective method of increase in resistance to corrosion-metallic cracking of the welded joints in the liquid ammonia medium.

At the same time it is necessary to perform the further investigations for more accurate definition of CMC resistance characteristics of steels, effect of residual welding stresses and thermal treatment on the process of initiation and propagation of cracks as well as execution of investigations to develop a welding technology providing a high uniformity of mechanical properties of butt-welded joints, first of all during realization of extreme actions and conditions.

References

[1] Pokhodnya I.K., Lebedev B.F. et al. Features of crack formation in the welded vessels for storage of liquid ammonia in the process of their service. *Automatic Welding*, 10, pp. 39-42, 1988.

[2] Basko E.M., Demygin N.E., Goncharova Y.E. Increase in safe operation life of vessels for storage of liquid ammonia under pressure. *Factory Laboratory*, 2, pp. 51-58, 1997.

[3] Basko E, Larionov V., Lazutin V. Influence of dynamic effects on the resistance of steel building structures to brittle failures. *WIT Transactions on the Built Environment*, 87, pp. 127–136, 2006.

[4] Kuzeev R.D. Experience of technical diagnostics (examination of industrial safety) of the spherical vessels with a volume of 2000 m^3 for storage of liquid ammonia at Mendeleevskazot Company Ltd., Republic of Tatarstan. *Research News*. Publication of Close Corporation "INOKhIM", 2-3, pp. 42-45, 2006.

[5] Karzov G.P., Leonov V.P., Timofeev B.T. Welded pressure vessels. "*Mashinostroyenie*", Leningrad, p. 287, 1982.

[6] Basko E.M. Cracking resistance of structural steels and construction units. *Proceeding Book of TsNIIproektstalkostruktsiya*. Research of brittle strength of the metal building structures, pp.112-128, 1982.

[7] Basko E.M., Makhutov N.A. Research of cracking resistance of structural steels under dynamic initiation and propagation of brittle cracks. Problems of fracture and safety of technical systems. *Collection of Scientific Proceeding*, Krasnoyarsk, pp. 44-49, 1997.

Early warning coordination centres: a systemic view

J. Santos-Reyes[1] & A. N. Beard[2]
[1]*Safety, Risk & Reliability Group, SEPI-ESIME, National Polytechnic Institute, Mexico*
[2]*Civil Engineering Section, Heriot-Watt University, Edinburgh, UK*

Abstract

Following the tsunami disaster in 2004, the General Secretary of the United Nations (UN) Kofi Annan called for a global early warning system for all hazards and for all communities. He also requested the ISDR (International Strategy fort Disaster Reduction) and its UN partners to conduct a global survey of capacities, gaps and opportunities in relation to early warning systems. The produced report, "Global survey of Early Warning Systems", concluded that there are many gaps and shortcomings and that much progress has been made on early warning systems and great capabilities are available around the world. However, it may be argued that an early warning system (EWS) may not be enough to prevent fatalities due to a natural hazard; i.e., it should be seen as part of a 'wider' or total system. Furthermore, an EWS may work very well when assessed individually but it is not clear whether it will contribute to accomplish the purpose of the total disaster management system; i.e., to prevent fatalities. There is a need for a systemic approach to early warning centres. Systemic means looking upon things as a system; systemic means seeing pattern and inter-relationship within a complex whole; i.e., to see events as products of the working of a system. A system may be defined as a whole which is made of parts and relationships. This paper proposes a preliminary model for an early warning coordination centre from a systemic point of view.
Keywords: risk, disaster, early warning, tsunami, systemic, SDMS, coordination centres, disaster management system.

WIT Transactions on Engineering Sciences, Vol 58, © 2007 WIT Press
www.witpress.com, ISSN 1743-3533 (on-line)
doi:10.2495/EN070121

1 Introduction

Natural disasters may be defined as events that are triggered by natural phenomena or natural hazards (e.g. earthquakes, hurricanes, floods, windstorms, landslides, volcanic eruptions and wildfires). Throughout history, natural disasters have exerted a heavy toll of death and suffering and are increasing alarmingly worldwide. During the past two decades they have killed millions of people worldwide, adversely affected the life of at least one billion more people. It has been estimated that the annual economic losses associated with such disasters averaged US $75.5 billion in the 1960s, US $138.4 billion in the 1970s, US $213.9 billion in the 1980s and US $659.9 billion in the 1990s [1].

In the late 1990s several natural disasters have occurred worldwide. However, on 26 December 2004 the biggest earthquake in 40 years occurred between the Australian and Eurasian plates in the Indian Ocean. The quake triggered a *tsunami* (i.e. a series of large waves) that spread thousands of kilometers over several hours. It is believed that several waves of the tsunami came at intervals of between five and 40 minutes. For instance, in Kalutara (a tourist resort in Sri Lanka) the water reached at least 1 Km inland, causing widespread destruction and death. The disaster left at least 165,000 people dead, more than half a million more were injured and up to 5 million others in need of basic services and at risk of deadly epidemics in a dozen Indian Ocean countries [2].

Following the tsunami disaster in 2004, the General Secretary of the United Nations (UN) Kofi Annan called for a global early warning system for all hazards and for all communities. He also requested the ISDR (International Strategy fort Disaster Reduction) and its UN partners to conduct a global survey of capacities, gaps and opportunities in relation to early warning systems. The produced report, "Global Survey of Early Warning Systems", concluded that there are many gaps and shortcomings and that much progress has been made on early warning systems and great capabilities are available around the world [3].

However, it may be argued that an early warning system may not be enough to prevent fatalities due to a natural hazard; i.e., it should be seen as part of a 'wider' system. Furthermore, an early warning system may work very well when assessed individually but it is not clear whether it will contribute to accomplish the purpose of the total disaster management system; i.e., to prevent fatalities. In other words, there is a need for a systemic approach and this will be discussed in the next section.

2 The need for a systemic approach

It has been argued that had a *tsunami* early warning system (EWS) been operational in the Indian Ocean, like the international tsunami warning system that covers the Pacific Ocean, the human toll might only have been a fraction of what it was [4]. However, it may be argued here that an early warning system should be seen as part of a 'wider system'; i.e. a total disaster management system. Furthermore, an early warning system may work very well when assessed individually but it is not clear whether it will contribute to accomplish the purpose of the total system; i.e. to prevent fatalities. For instance, a regional EWS may only work if it is well co-ordinated with the local warning and emergency response systems that ensure that the warning is received, communicated and acted upon by the potentially affected communities. It may be argued that without these local measures being in place, a regional EWS will have little impact in saving lives. Researchers argued that unless people are warned in remote areas, the technology is useless; for instance McGuire [5] argues that:

"I have no doubt that the technical element of the warning system will work very well,"..."But there has to be an effective and efficient communications cascade from the warning centre to the fisherman on the beach and his family and the bar owners."

Similarly, McFadden [6] states that:

"There's no point in spending all the money on a fancy monitoring and a fancy analysis system unless we can make sure the infrastructure for the broadcast system is there,"... "That's going to require a lot of work. If it's a tsunami, you've got to get it down to the last Joe on the beach. This is the stuff that is really very hard."

Given the above, the paper argues that there is a need for a systemic approach to early warning centres. *Systemic* means looking upon things as a system; *systemic* means seeing pattern and inter-relationship within a complex whole; i.e., to see events as products of the working of a system. *System* may be defined as a whole which is made of parts and relationships. Given this, 'failure' may be seen as the product of a system and, within that, see death/injury/property loss etc. as results of the working of systems. This paper proposes a preliminary model of early warning coordination centres from a systemic point of view.

3 Early warning coordination centres

A Systemic Disaster Management System (SDMS) model has bee constructed by adopting a systemic approach and this will be described briefly in section 3.1. Section 3.2 describes a preliminary model for an early warning coordination centre which is seen as part of the SDMS model.

3.1 A Systemic Disaster Management System (SDMS) model

The SDMS model is intended to maintain disaster risk within an acceptable range in an organization's operations in relation to natural disaster management. The model is proposed as a *sufficient* structure for an effective disaster management system. It has a fundamentally *preventive* potentiality in that if all the subsystems and connections are present and working effectively the probability of a failure should be less than otherwise. Table 1 summarises the fundamental characteristics of the SDMS model.

Table 1: SDMS' characteristics.

1	The SDMS & its 'environment'
2	A recursive structure (i.e. 'layered') and relative autonomy
3	A structural organization which consists of a 'basic unit' in which it is necessary to achieve five functions associated with systems 1 to 5. (See Figure1). (a) system 1: disaster-policy implementation (b) system 2: disaster- total early warning coordination centre (TEWCC) (c) system 2*: disaster-local early warning coordination centre (LEWC) (d) system 3: disaster-functional (e) system 3*: disaster-audit (f) system 4: disaster-development (g) system 4*: disaster-confidential reporting system (h) system 5: disaster-policy Note: whenever a line appears in Figure 1 representing the SDMS model, it represents a channel of communication.
4	The concept of MRA (Maximum Risk Acceptable), Viability and acceptable range of risk.
5	Four principles of organization
6	'Paradigms' which are intended to act as 'templates' giving essential features for effective communication and control.

See Beard [7] and Santos-Reyes and Beard [8] for details of the origin and development of the model; a full account of the above characteristics is described in [9,10]. A brief description of the structural organization of the model will be given in the subsequent paragraphs.

3.1.1 A structural organization which consists of a 'basic unit' (i.e. systems 1-5)

(a) System 1: Disaster- policy implementation, implements safety policies in the organization's operations. System 1 consists of one or more operations (e.g. disaster operations at the level of a country, or zone or region).

(b) System 2: Disaster-TEWCC, coordinates all the activities of the operations that form part of system 1 (see Figure 1) and in relation to the 'total environment'. Furthermore, it also coordinates other local early warning coordination centres (LEWCCs). System 2 along with system 1, implements the safety plans received from system 3.

(c) System 2*: Disaster-LEWCC, is part of system 2 and it is responsible for communicating advance warnings to other early warning coordination centres and to key decision makers in order to take appropriate actions prior to the occurrence of a major natural hazard event. (See section 3.2 for details about this).

(d) System 3: Disaster-functional is directly responsible for maintaining disaster risk within an acceptable range in system 1 on a daily basis. It ensures that system 1 implements the organization's safety policy.

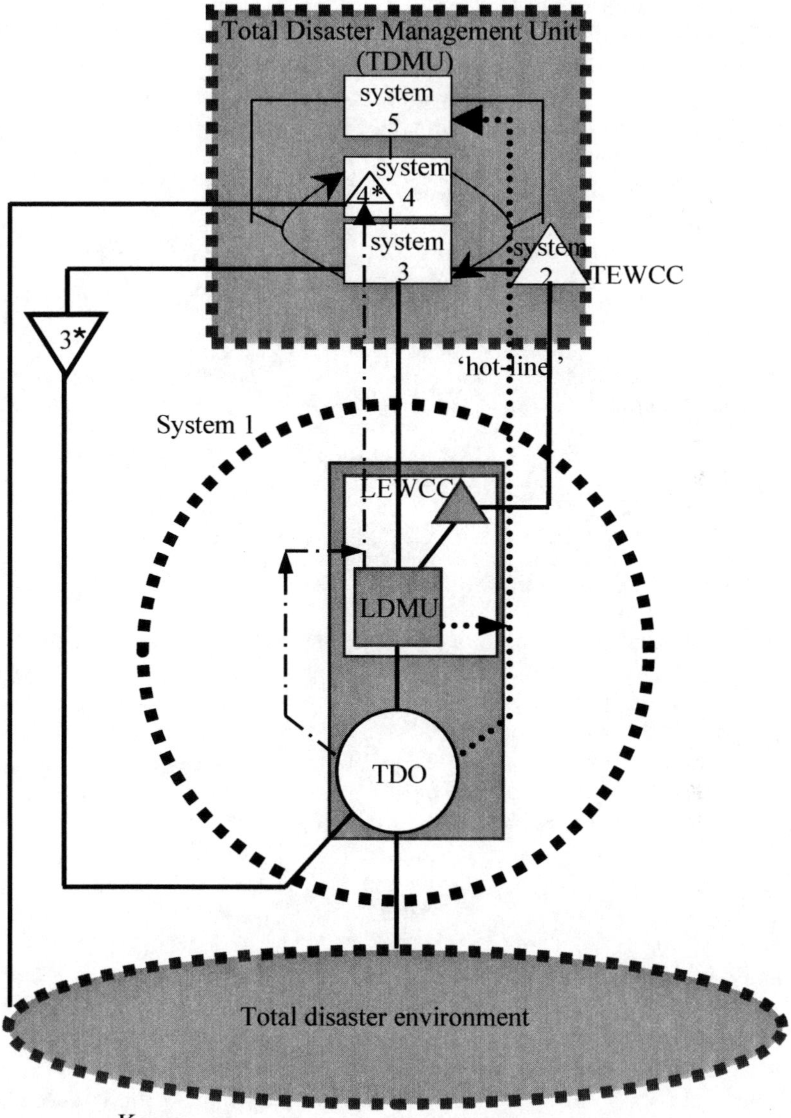

Key:
TDMU = Total Disaster Management Unit
TEWCC = Total Early Warning Coordination Centre
LEWCC = Local Early Warning Coordination Centre
LDMU = Local Disaster Management Unit
TDO = Total Disaster Operations

Figure 1: A Systemic Disaster Management System (SDMS) model.

(e) System 3*: Disaster-audit, is part of system 3 and its function is to conduct audits sporadically into the operations of system 1. System 3* intervenes in the operations of system 1 according to the safety plans received from system 3.
(f) System 4: Disaster-development, is generally concerned with the 'total environment' and its function is to conduct research and development (R&D) for the continual adaptation of the organization. By considering strengths, weaknesses, threats and opportunities, system 4 can suggest changes to the organization's safety policies.
(g) System 4*: Disaster-confidential reporting system, is part of system 4 and it is concerned with confidential reports or causes of concern from any person of the public about any aspects, some of which may require the direct intervention of system 5.
(h) System 5: Disaster-policy, is responsible for deliberating safety policies and for making strategic decisions. System 5 also monitors the activities of system 4 and system 3.
(i) 'Hot-line': Figure 1 shows a dash line directly from system 1 to system 5, representing a direct communication or 'hot-line' for use in exceptional circumstances; for example, during an emergency.

3.2 Early warning coordination centres

The function of system 2 is to coordinate the activities of the operations of system 1. To achieve the plans of system 3 and the needs of system 1, system 2 gathers and manages the safety information of system 1's operations.

In a relatively well coordinated system the information flows might be according to the arrangement shown in Figure.2. In general, the arrangement indicates that if a deviation occurs from the accepted criteria, then the functions of the LEWCC within system 1 are the following:

Firstly, detect any deviation from the accepted criteria (see action point '2' in Table 2 and Figure 2).

Secondly, issue the disaster warning simultaneously to:
(a) LDMU; so that, it implements the pre-planned 'measures' in the operations; see action points '3', '4' and '5' (e.g. evacuation, search and rescue, emergency medical services).
(b) other LEWCCs through action point '2A'. Similarly, these coordination centres have to assess consequences and implement measures within their operations and make reports quickly to system 2 (TEWCC); see Table 2 & Figure 2.
(c) System 2 (TEWCC) through action point '4A'. By receiving the warning it takes fast corrective action, either through the channels of communication that connects the LEWCC or via system 3 and this is shown in Figure 2. Some of the functions of the TEWCC are: collection and compilation of information from the affected area, supply of information to System 3.

Figure 3 shows a preliminary model for EWCCs at National and Regional levels.

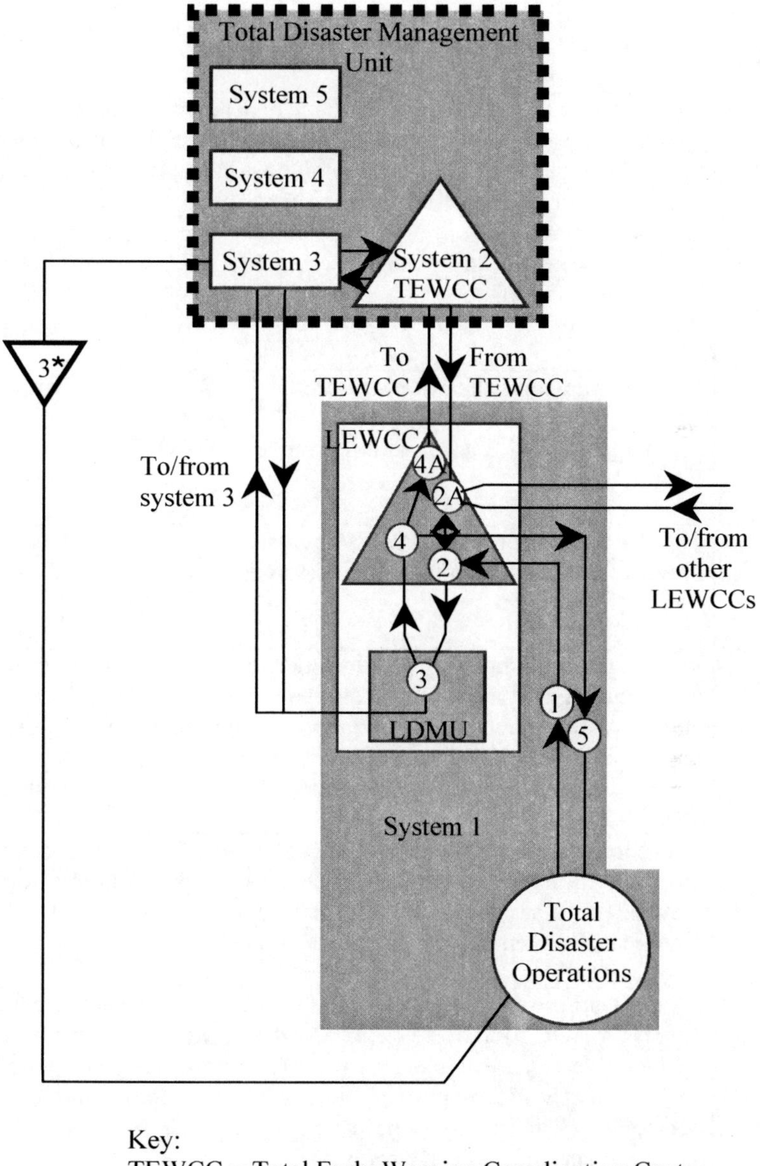

Key:
TEWCC = Total Early Warning Coordination Centre
LEWCC = Local Early Warning Coordination Centre
LDMU = Local Disaster Management Unit

Figure 2: Early warning coordination centres – a preliminary model.

Table 2: 'Action points' - Figure 2.

Action point	Description
1	Flow of data related to any particular sensor system (e.g. those related to an earthquake/tsunami: ocean bottom pressure sensors, buoys, tide gauges, etc.). The communications may be done via wire line, wireless, satellite, etc.
2	Comparison/analysis of the data/information being received. If any deviation from the pre-planned acceptable criteria occurs then it issues the warning to action points '2A' and '3' as indicated in Figure 2.
2A	(a) Communicates the warning to other LEWCCs (see Figures 2 and 3). (b) It also receives information from the TEWCC as shown in Figure 2.
3	The function of the LDMU is to respond to the warning and prevent fatalities due to the natural hazard.
4	Planning and taking measures in order to respond to the warning. For example, preparedness for any emergency, in particular those, which strike without notice, requires a plan; some of the aspects that should be taken into account may be: the identification of possible emergency situations which may occur in a particular area, etc.
4A	Issues the warning to the TEWCC (See Figure 2) and by receiving all this information, the TEWCC enables to take a 'higher' order view of the total consequences. It will report to system 3, which is on the vertical command channel (see Figures 1 and 2).
5	(a) the warning is issued. The public may be informed by public address systems, radio, TV, etc. At present, no early warning can be given for an earthquake. However, some of the conventional ways of early warning may be used; for example, an erratic behaviour of animals just before an earthquake have been used since ancient times as early warning for such events. (b)} Implementation of pre-planned 'measures' to evacuate safely and prevent fatalities due to natural hazards; e.g. provision of medical services, search, rescue and evacuation. The primary concern should be the safety of the public; protection of property by the police and fire fighters.

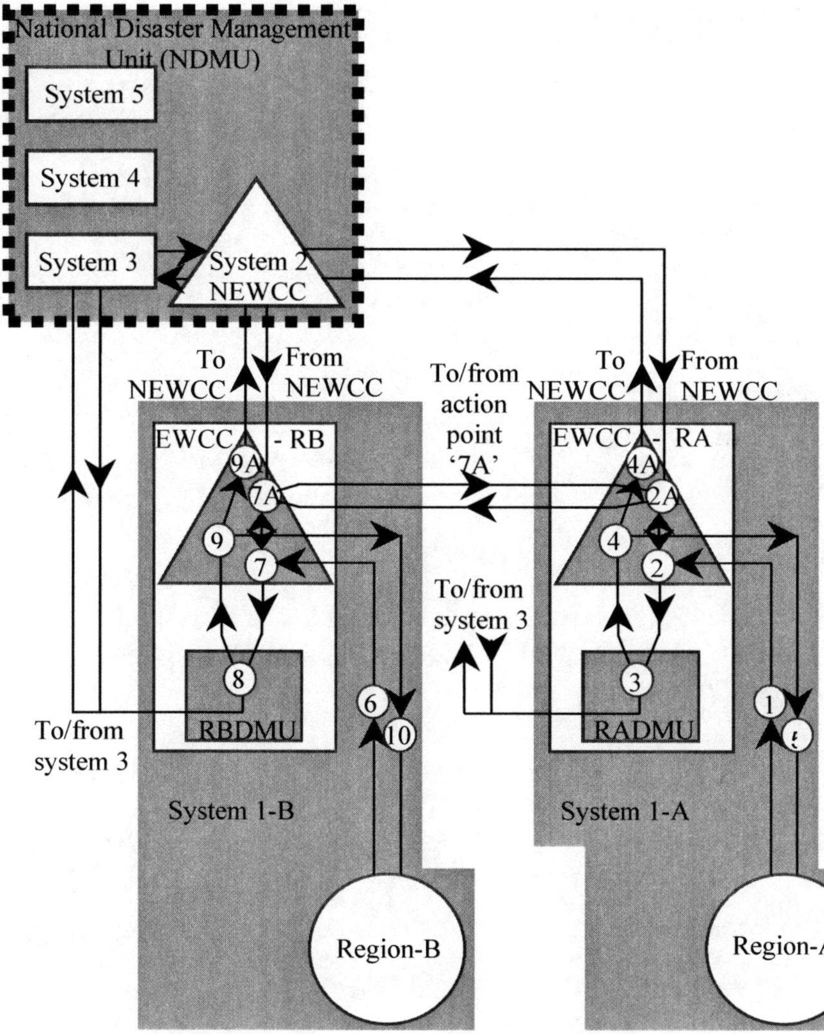

Key:
NDMU = National Disaster Management Unit
NEWCC = National Early Warning Coordination Centre
EWCC-RA = Early Warning Coordination Centre – Region A
EWCC-RB = Early Warning Coordination Centre – Region B
RADMU = Region A – Disaster Management Unit
RBDMU = Region B – Disaster Management Unit

Figure 3: Early warning coordination centres – at national and regional levels.

4 Conclusions and future work

A preliminary model for an Early Warning Coordination Centre (EWCC) has been put forward. The EWCC is associated with system 2 of a Systemic Disaster Management System (SDMS) model which has been constructed by using the concepts of systems. Further research is needed in order to construct recursive early warning coordination centres; i.e., from, international to national, regional to community level from a *systemic* point of view. It is hoped that this approach will help to provide "…an effective and efficient communications cascade from the warning centre to the fisherman on the beach & his family and the bar owners" [5].

Acknowledgements

This project was funded by CONACYT & SIP-IPN under the following grants: CONACYT: No-52914 & SIP-IPN: No-20071593.

References

[1] UNDP. Reducing Disaster Risk- a challenge for development. United Nations Development Programme (UNDP), 2004, Online. www.undp.org/bcpr.

[2] BBC. Tsunami disaster. Online. URL:http://news.bbc.co.uk/go/pr/fr/-/1/hi/world/asia-pacific/4136289.stm

[3] Annan, K. In the Secretary-General's report "In larger Freedom". Report A/59/2005, paragraph 66, 2005.

[4] UNDP. Survivors of the tsunami: One year later. United Nations Development Programme (UNDP), 2005, Online. www.undp.org/bcpr.

[5] Kettlewell, J. (2005). Early warning technology – is it enough? Online. http://news.bbc.co.uk/go/pr/fr/-/2/hi/science/nature/4149201.stm.

[6] Kettlewell, J. Tsunami alert technology – the iron link. Online. http://news.bbc.co.uk/go/pr/fr/-/2/hi/science/nature/4373333.stm.

[7] Beard, A. N. Some ideas on a systemic approach. *Civil Engineering & Environmental Systems*, 16, pp. 197-209, 1999.

[8] Santos-Reyes, J., and Beard, A. N. A systemic approach to fire safety management, *Fire Safety Journal*, 36, pp. 359-390, 2001.

[9] Santos-Reyes, J. (2006). A systemic approach to disaster management. *Proc. Of the International Symposium on Management System for Disaster Prevention*, ISMD, 9-11 March, Kochi, Japan, 2006.

[10] Santos-Reyes, J. & Beard, A. N. A systemic approach to managing natural disasters. Forthcoming paper in *Natural Hazards* journal.

Author Index